Design Representation

Springer
London
Berlin
Heidelberg
New York
Hong Kong
Milan
Paris
Tokyo

Gabriela Goldschmidt and
William L. Porter (Eds.)

Design
Representation

With 95 Figures

 Springer

Gabriela Goldschmidt, M.Arch
Technion-Israel Institute of Technology, Haifa 32000, Israel

William L. Porter, PhD, FAIA
Massachusetts Institute of Technology, Cambridge, MA 02139, USA

Cover illustration: Leonardo da Vinci, study sketches for a new Palazzo Medici in Florence, c. 1515. Copyright Biblioteca Ambrosiana - Authorization Number F 145/03.

ISBN 978-1-84996-904-8   e-ISBN 978-1-85233-863-3

British Library Cataloguing in Publication Data
Design representation
  1. Design, Industrial 2. Architectural design
  I. Goldschmidt, Gabriela II. Porter, William L.
  745.2

Library of Congress Cataloging-in-Publication Data
International Design Thinking Research Symposium (4th: 1999: Cambridge, Mass.)
  Design representation/Gabriela Goldschmidt and William L. Porter, eds.
    p.cm.
  Includes bibliographical references.

  1. Communication in architectural design–Congresses. 2. Communication in engineering design–Congresses. 3. Communication in design–Congresses. 4. Design–Research–Congresses. I. Goldschmidt, Gabriela, 1942–   II. Porter, William L., 1934–   III. Title.
NA2750.I6 1999
729′.01–dc21   2003054436

Springer-Verlag London Berlin Heidelberg
a member of BertelsmannSpringer Science+Business Media GmbH
http://www.springer.co.uk

Preface

In April 1999, the undersigned co-chaired a meeting entitled "The 4th International Design Thinking Research Symposium," or DTRS '99, at the Massachusetts Institute of Technology (MIT). The theme of the symposium was *Design Representation*. The present book has its roots in that event.

We have been interested in design representation for a long time, in ways both similar and complementary. We have for years written and taught courses that explored design representation, although we used other terms to describe what we were looking at. We learned that other people, in various design and design research domains, were showing increasing interest in questions pertaining to representation. Therefore, we chose this theme as the topic of the 4th Design Thinking Research Symposium when our good fortune destined us to organize and chair it. The journey we have undertaken, starting with the inception of the idea for the symposium, came to a conclusion only once these pages were assembled. It was a true intellectual adventure that we enjoyed tremendously, as it gave us an opportunity to learn more, to ask many questions, and to create a fruitful dialogue with contributors to this book. Although all the authors attended the DTRS meeting, and most of the chapters in this book build on presentations made there, the book is in no way a replication of the meeting's proceedings.[1] Nor does it resemble two special issues of professional journals, guest-edited by us, that feature DTRS'99 papers.[2]

Deciding on the precise focus, the structure and the layout of this book was no easy task, even after many months of dealing with the topic. We approached the job of editing this book like a design job and we allowed the material to talk back to us. Our deepest appreciation is herewith extended to the authors who shared their thinking with us, who responded to our questions, who made fine suggestions, and who were very patient with us. We are likewise extremely obliged to the entire *Design Thinking Research* community, whose many members so enthusiastically responded to the idea of holding the meeting in 1999. We believe that the clear voices of approval that have come out of this community made it possible for us to target Design Representation as the focal point of our work. We are indebted to the Department of Architecture at MIT for the support – both intellectual and material – that it has provided to the DTRS meeting and its preparation. Major support from Autodesk, Inc., auto•des•sys, Inc., and the Microsoft

---

1. The pre-proceedings of DTRS'99 included 50 papers, all of which were presented at the meeting. The papers were accepted following reviews of 150 abstracts that had been submitted.
2. Design Studies 21:5, 2000, and Automation In Construction 10:6, 2001.

Corporation helped make for a richer event. A considerable number of MIT students and employees worked hard to make it a successful experience. Without the help of all of them, we would not have had the base that paved the way for this book. Infinite gratitude goes to the Graham Foundation for Advanced Studies in the Fine Arts, whose generous grant enabled us to bridge the geographic distance that separates us and to collaborate on our project in ways that would have not been possible otherwise. Finally, we are most grateful to Pamela Siska for considerably uplifting the language and form of our texts, and to Francesca Warren, our editor at Springer Verlag, whose enthusiasm and faith in this endeavour were decisive in bringing it to fruition.

Gabriela Goldschmidt, Haifa
William L. Porter, Cambridge, Massachusetts

Contents

List of Contributors ix

Introduction xi
Gabriela Goldschmidt and William L. Porter

Part I From the Perspective of Architecture

1 Distance and Depth 3
   Penny Yates

2 Graphic Representation as Reconstructive Memory: Stirling's
   German Museum Projects 37
   Gabriela Goldschmidt and Ekaterina Klevitsky

3 Designers' Objects 63
   William L. Porter

Part II From the Perspective of Engineering

4 Distributed Cognition in Engineering Design: Negotiating
   between Abstract and Material Representations 83
   Margot Brereton

5 Design Representations in Critical Situations of Product
   Development 105
   Petra Badke-Schaub and Eckart Frankenberger

6 Impromptu Prototyping and Artefacting: Representing Design
   Ideas through Things at Hand, Actions, and Talk 127
   Gilbert D. Logan and David F. Radcliffe

Part III **Beyond Disciplinary Perspectives**

7 Cognitive Catalysis: Sketches for a Time-lagged Brain 151
   Jonathan Fish

8 The Thoughtful Mark Maker – Representational Design Skills
   in the Post-information Age 185
   Martin Woolley

9 Design Representation: Private Process, Public Image 203
   Gabriela Goldschmidt

Index 218

Contributors

Petra Badke-Schaub
  Department of Psychology
  and Methodology, University
  of Bamberg, Bamberg,
  Germany
  petra.badke-schaub@
  ppp.uni-bamberg.de

Margot Brereton
  School of Information
  Technology and Electrical
  Engineering, University of
  Queensland, Brisbane,
  Australia
  margot@itee.uq.edu.au

Jonathan Fish
  Independent Researcher,
  Le Gres Frausseilles, Cordes,
  France
  jcfish@easynet.fr

Eckart Frankenberger
  Research and Technology
  and Future Projects,
  Airbus Deutschland GmbH,
  Hamburg,
  Germany
  eckart.frankenberger@airbus.
  com

Gabriela Goldschmidt
  Faculty of Architecture and
  Town Planning, Technion,
  Haifa, Israel
  gabig@tx.technion.ac.il

Ekaterina Klevitsky
  Faculty of Architecture and
  Town Planning, Technion,
  Haifa, Israel
  arekater@tx.technion.ac.il

Gilbert D. Logan
  Rehabilitation Engineering
  Centre, Royal Brisbane Hospital,
  Queensland, Australia
  gilbert_logan@health.qld.gov.au

William L. Porter
  Department of Architecture,
  Massachusetts Institute of
  Technology, Cambridge, MA,
  USA
  wlporter@mit.edu

David F. Radcliffe
  Catalyst Centre, School of
  Engineering, University of
  Queensland, Brisbane, Australia
  d.radcliffe@uq.edu.au

Martin Woolley
  Department of Design Studies,
  Goldsmiths College, University
  of London, London, UK
  martin@mwoolley.demon.co.uk

Penny Yates
  New Jersey School of
  Architecture at New Jersey
  Institute of Technology, NJ, USA
  pyates2724@aol.com

Introduction

There can be no design activity without representation. Ideas must be represented if they are to be shared with others, even shared with oneself! Different representational modes and strategies afford distinctive opportunities for reading or for transforming design ideas. We believe Design Thinking Research must address these and other issues of representation as well as the underlying theories.

The above quotation is from the brief text that accompanied the announcement of the Design Thinking Research Symposium on *Design Representation* in 1999. We believed then, and we still believe now, that the notion of design representation is more complex, more serious, and more important than has been acknowledged in contemporary research and scholarly writings. The scope of Design Representation is formidable and to attempt comprehensive treatment of this theme would not be a reasonable mission. Rather, we thought it a good idea to aim at introducing the issues that we find most fascinating by illuminating them from different angles and perspectives. This has been our strategy for this book.

The open questions that we have posed regarding representation commence with the most basic dilemma: what are representations? Should we regard everything as representation, as cognitive science – or at least some of its varieties – has done? Are representations to be seen as interim entities that always stand for something else, which is the real reason for evoking them? Or, conversely, are representations solid realities, objects which, once generated, have a life of their own regardless of their functions as similes or simulations of the "real" thing? Is the difference sharp and impermeable, or can object and representation be merged, or be transformed into each other? If so, is this a reversible process? And what form do representations (and, alternatively, objects) take? Abstract or material? Propositional or pictorial? Should internal representations, in the privacy of one's own mind, be included in an exploration of design representation? To what extent are normative representations personal? Do all representations obey common rules of production and why are some of them considered idiosyncratic? What are the historical, cultural and social dimensions of representations? Can they be clearly stated? As we shall see, the authors of the chapters in this book have included language, drawings of various types, and objects as belonging in the inventory of representational forms used in design. Is the choice of representational form a matter of personal preference? Is it dictated by the task or might it be otherwise contextually dependent? Many of the chapters tackle these questions.

Finally, there is the often-overlooked question of viewing representations. How are representations viewed and by whom? Can a representation be cognized without the full integration of the viewer into the resulting image?

And who is the viewer – is he or she a "subjective" viewer whose position in space and way of looking at something are instrumental to the representation? Or can he or she be a disembodied observer who receives the representation without interacting with it and without having any effect on its nature? In fact, do we always have a way of knowing who the observer is or will be in the future? If the observer or viewer is not known and not predictable, can we surmise his or her reading of the representation? We recognize that representations may be more or less successful in conveying messages – intended and unintended ones. As makers of representations, do we know how the "other" is likely to interpret our representations? Do we depend on worlds of shared notions, conventions, notations, symbols, and values? None of these questions are easily answered and, indeed, this book does not pretend to offer answers. Instead, as stated above, the various chapters present these questions in a variety of different ways that enrich our understanding of the notion of representation.

So far, not much has been said about design. After all, our study of representation pertains to design and is anchored in the context of design thinking research. Such research often remains within the realm of one particular design discipline, such as architecture, engineering, or industrial design. Studies of the kind we have undertaken often stress the distinctiveness of practices, beliefs, and norms within a discipline, assuming that the uniqueness of each field is sufficiently strong to warrant a within-domain enquiry. Alternatively, one could become interested in crossing the boundaries between particular disciplines and establishing the commonalties that tie all design disciplines together into an art or science of design. In such cases, representations are treated in accordance with their universal roles in supporting the design process towards the making of something new. We have chosen to organize the chapters in this book along disciplinary lines because we believe that important differences among design disciplines do exist and should by no means be ignored or overlooked. Architecture and mechanical engineering were selected as two poles of the disciplinary continuum, in which such differences can be played out easily. At the same time, we agree that there is a level at which design is fundamentally one and the same activity across fields and domains. We want to stress that which is shared alongside that which is better dealt with separately. Accordingly, the book is divided into three parts that highlight the different perspectives we have chosen: the perspective of architecture, the perspective of engineering, and, finally, a view from beyond disciplinary perspectives.

From the Perspective of Architecture

Representations are necessary for the practice of architectural design. They take the form of drawings of many kinds, three-dimensional models, and nowadays, of course, a variety of digitally based images. Designers must work with representations in the production of architectural works, as in all but very few situations it is impossible to assess design proposals without first employing media other than full-scale structures. Therefore, working with reduced-scale representations has always been habitual in architecture. However, many questions arise regarding the nature of architectural representation. What do they mean to the designer or to the eventual user or observer? How do designers interact with them as they design? To what extent can they embody subjectivity, effect, or even intellect? Are they merely representative of the building to be ultimately built, or are they final products in their own right? Can they express certain ideas with greater force than the buildings they presumably represent? What inspires the choice of representational modes and media? When and why are new types of representation introduced? These are some of the questions that the three following chapters explore in very different ways.

According to her own testimony, Penny Yates in Chapter 1 undertakes a "search for the presence of the subjective viewer in design representation." She leads this search by analyzing artefacts rather than speculating about their role in the design process, although she does refer to architects' intentions when they are explicitly known and documented, as is the case of some of Le Corbusier's projects. Yates distinguishes between object-centred representations, which deal well with distance, and subject-centred representations, which are more concerned with depth. Depth, a more profound way of experiencing spatial relations than distance, is what the "perceiving subject" feels, whereas distance is the way in which the "disembodied observer" would describe his or her experiences with the architectural object. Yates uses a large number of carefully selected examples to make and substantiate her arguments. In a long discussion she lays out the difference between one- and two-point perspective, stressing the crucial importance of the viewer's station point to the three-dimensional perceptions that may be expected. The viewer's position is also important to the question of symmetry versus asymmetry of representation, a factor that strongly impacts experience – for example, when reading depth off a series of planes in an enfilade of spaces. Yates hints at an inadequacy of computational representation applications which, to date, engage the object and do not depart from what the human subject might sense in a delimited space, as well as outside of it.

In Chapter 2, Gabriela Goldschmidt and Ekaterina Klevitsky expand on what Yates has chosen to leave out: the intentions of the architect as revealed through representations in professional publications. The case they look at in detail is the three design competitions for German museums that the British architect James Stirling and his partners worked on and published in the mid-1970s. The publications included, in addition to conventional drawings, also preliminary sketches and highly abstract down and up axonometric views, which, according to Goldschmidt and Klevitsky, were added for the purpose of elucidating the central design concept. The authors argue that the unusual (for its time) publication of sketches and analytic "axos" signalled a new approach to public architectural representation in the era of postmodernism. The architects, who wanted to present ideas and not just transmit factual depictions of the buildings they designed, offered a pictorial narrative that told the story of these designs that, in this case, was largely the story of public paths and spaces that dominated all three schemes. The analysis presented in Chapter 2 concerns the design concepts evoked by the architects and the pictorial means used to convey them as a narrative that portrays the trajectory from initial thoughts to final plans, with commentary by way of abstracted three-dimensional views. Goldschmidt and Klevitsky conclude that a shift in interest from design product toward design process has led these postmodern architects to reconstruct for the public a memory of the values with which the conception of these projects was invested.

The last chapter in Part I is by William Porter. In Chapter 3, Porter explores how designers establish a discourse between themselves and the objects they are contemplating. Through the use of several examples, he explores how that discourse informs their understanding of objects that are integral to their design process, as well as those that fall outside that process. These objects include what we customarily refer to as "representations," as well as those we term "objects." The mode of interaction, he argues, is the same. Given that recognition, objects made for the purposes of design, commonly thought of as "representations," take on the full significance for the user of any object, whether a means to a design end or not. Thus designers' objects embody, symbolize, and mean in ways that are identical to the cultural artefacts we identify as buildings or paintings or other "finished" works. In their more highly charged role, it is easier to see how they can interact productively with the designer during the course of design. Indeed, many of these "designer objects," like the extraordinary sketches of Ludwig Mies van der Rohe or Alvar Aalto and many other great architects, have taken their place as fine works in and of themselves. Porter has chosen to explore these ideas through a variety of episodes drawn from his own experience and that of his students. These include explorations through the voice of the other, through the experience of place and building, through conjectural and reconstructive exercises to understand specific objects, and through the playing of games. Objects may be created that are not integral to the production of the building (or other design), yet are integral to the expression of ideas having to do with it. It is the expressive content of these objects, as well as their representational link to the building related to them, that makes them valuable to the designer as well as to others. It is specifically the nature of the discourse with these objects that will determine the strength of the linkage between them and the designer's experience, skills, memory, and powers of empathy.

From the Perspective of Engineering

Mechanical engineering design activities are normally carried out by design teams and not by individual designers. There is a consensus regarding the significance of communication among team members to a fruitful design process: communication is aimed at providing information, presenting, assessing, refining and challenging ideas, and representing design queries and decisions, tentative as well as final. Researchers of engineering design processes stress the notion that communication is a complex operation that involves language, gestures, graphic representation (in the form of drawings) and material objects of various kinds. Different settings may, of course, prioritize different modes of communication among design team members. The type of task, the experience of the designers, and the social relations among them are but some of the factors that may affect communication. The chapters in Part II present studies of design representation in three different settings. First, an educational environment in which students learn how to design (learning by doing); second, in an organizational setting – a typical instance of work in industry; and third, in a specialized case of custom adaptation of engineering solutions to specific, individual needs. Taken together, these papers portray the intricacies and complexities of representation in engineering design, and stress the need to understand and encourage the use of multi-modal means of representation.

The study by Margot Brereton in Chapter 4 is set in the studio where student engineering designers work on a number of design assignments. The study demonstrates that learning occurs through continually challenging abstract representations against material representations. The gaps between the two modes of representation inspire further design activity: representation in the two modes informs and advances the design solution, enhances the designers' understanding of design requirements, and brings to light implicit design assumptions. Hardware repertoires are extended, and fundamental engineering concepts are sorted out through a continual process of representation and rerepresentation in abstract terms and in material form. In particular, the role of material representation in supporting cognitive activity is instantiated through rich examples concerning design exercises such as the design of a crane or a kitchen scale. In these examples hardware is shown to assume a variety of roles in mediating the learning process, including those of starting point, thinking prop, medium of integration, embodiment of abstract concepts, and more. Brereton's in-depth analysis is grounded in a theory of the primacy of multi-modal representational modes in learning, and in particular negotiations between abstract and material representations. In our estimation this theory can and should be extended beyond the scope of the educational setting, into the realm of practice in general, where ill-structured problems habitually require relearning and reformulation of problems.

In Chapter 5, Petra Badke-Schaub and Eckart Frankenberger are concerned with the availability of information through communication among team members who are involved in the development of a new product. Information is transmitted through representations of various kinds: verbal, written, sketches, drawings, and electronic data. The present study concentrates on verbal information transfer in what the authors call "critical situations" (defined by task requirements) of the design process. Data was collected by observing design activity on a daily basis for a relatively long period, using three criteria: "individual prerequisites," "prerequisites of the group," and "external conditions." To those we must add the givens of "the task": in one case the redesign of a pneumatic fruit press, and in another case the developing and redesigning of several components of a particleboard production plant. The study found that the main venue for information transfer in teamwork (critical situations) was verbal exchanges. Designers testified that asking colleagues was their preferred way of acquiring specifically required information, and that informal conversations also provided very useful information. Further analysis showed that critical evaluations are mostly achieved through positive affirmations, but not exclusively so. "Positive affirmation" representations are particularly instrumental in enhancing a good group climate, which in turn contributes primarily to activity of the type "solution search."

The third chapter in Part II, Chapter 6 by Gilbert Logan and David Radcliffe, is dedicated to a unique engineering task in which a team works to adjust and refine engineering solutions to the personal needs of patients in a Rehabilitation Engineering Centre. A case in point is the seating clinic, where a patient born with congenital amputations of his arms and his right leg is seeking help in adapting his wheelchair to the operation of a laptop computer, using his degenerated left leg. By the nature of the task, team members continually communicate with each other and with the patient. The authors analyze communication using three categories of representation: talk, action (such as gestures and mimicry), and the use of artefacts. The combination of the three factors in design activity is called "artefacting" and the act of using objects at hand to simulate design ideas is called "impromptu prototyping." Sessions of routine work in the clinic were videotaped and later parsed into "events". A fine-grained analysis of these events shows how interrelated the three types of representation (divided into subcategories) are; in fact, in more than half of the events, all three types were detected, hence an emphasis on impromptu prototyping. We believe that this analysis mirrors a large number of engineering design episodes in which goal-oriented behaviour takes advantage of all the available means to reach the best possible solutions, relying primarily on common sense and on affordances provided by contextual settings.

Beyond Disciplinary Perspectives

Every design discipline has developed its own traditions, norms, and conventions of representation, commensurate with its evolving culture(s), professional objectives, and the organization of the workplace and work methods. Architecture, for example, has been primarily concerned with space and its enclosure, with questions regarded as pertaining to "aesthetics", with cultural integrity and continuity over time, and hosts of other material, as well as non-material, mostly qualitative, issues. Engineering design is much more, if not exclusively so, about material qualities of objects. Function and performance precede consideration of any independent aesthetic nature, and design entities are not single, individually designed "one-off" products but often "revised models" of prior existing products, the properties of which are usually quantitatively evaluated. Representations in the two fields should, and do, reflect these differences. They are embedded in their respective cultures and respond primarily to the needs and expectations of members of their respective professional communities and their audiences. In no way do we wish either to overlook the differences or mitigate their significance. However, at a fundamental level, there also exist considerable commonalties in design thinking and therefore also in representational properties and in the way that designers in all disciplines go about generating and utilizing them. Commonalities are discernible when assertions proclaimed in the context of one discipline resonate in the context of another discipline as well. For example, when Brereton talks about negotiations between abstract and concrete representations in engineering design, we can easily map her descriptions on to architectural design. Likewise, the dichotomy between the role of representations as simulations and as objects in their own right that Porter brings up in the context of architecture is not foreign to engineering. Part III contains three chapters that address issues of universal significance to all design disciplines: the role of sketching in design thinking, the evolution of representational design skills, and, finally, a possible paradigm for the study of design representation.

Jonathan Fish in Chapter 7 is interested in the role of sketching in solving problems that require visual invention, as is typically the case in the various design domains. Sketching, he claims, amplifies the mind's ability to translate descriptive ideas to depictive images and vice versa. Such back-and-forth translations are essential in complex visual invention processes because descriptive thought and depictive mental images are represented in different components of our working memory: the linguistic and visuo-spatial working memories, respectively. Fish draws on a strong metaphor – that of chemical catalysis – to describe how sketching affects the two "reactants" – the descriptive and depictive thoughts, and images. He proposes that we are in need of an amplifying device because our mental resources for visual invention, as embodied in visual mental imagery, were originally better adapted for perceiving and acting on the immediate present, than for imagining the future. Our hominid ancestors' survival depended on the brain's capacity, with the help of mental imagery, to make fast, flexible responses to unexpected opportunities and dangers, and not on its ability to plan for distant futures. Designing falls into the category of planning for the future, of course, and evolution has not adapted our brain's imagistic capacities fast enough to deal successfully with such problems. Therefore, external help in the form of sketches as an amplification device or catalyst plays such an important role in design endeavours. Fish explores his metaphor with reference to Robert Welch's sketches for a stainless-steel serving collection.

A very different vista on design is cast by Martin Woolley who looks at the implications of the powerful impact of computing on design representation skills in Chapter 8. Woolley asserts that professional expertise depends on a command of state of the art skills. Emerging Information Technologies are perceived as replacements for traditional skills, and therefore as potentially threatening to deskill professionals. The problem that Woolley formulates is that of reskilling. He discusses traditional design representation skills from the point of view of the control of tools used to exercise skilful practice. In this view the development of skills is intimately related to the control of tools. The present Information Age is defined as one of "self-directed" skills in which the autonomous designer/maker uses learned skills that are redirected under individual control. Controlling the use of tools and skills could be insufficient because the tools themselves have properties over which the designer has no control and the result might be deskilling. To overcome the problem of deskilling, Woolley proposes that the Post Information Age be one of "self-originated" skills in whose generation, development, and ownership the designer is to be proactive. This activity would empower the design practitioner and ensure his or her control not only over the use of tools, but also over the relevancy and appropriateness of the properties and capabilities they incorporate.

Finally, in Chapter 9 Gabriela Goldschmidt seeks to conceptualize an epistemological framework for the study of design representation. She proposes two perpendicular axes of enquiry. The first observes representation along the dimensions of cognition, history and culture, and technology and media. Fish's chapter, for example, would be considered as addressing the concerns of the cognitive dimension, whereas Woolley's contribution clearly falls within the technology and media dimension. Other chapters in this book, such as that by Penny Yates, fit in the history and culture dimension. The second axis of enquiry is that of the public versus the private contexts of representation which, in this view, are motivated by different goals and aspirations, although the underlying representational norms and conventions used in both may be similar. A clear distinction between the two is drawn, as instantiated by Goldschmidt and Klevitsky in Chapter 2. The two axes are conceived by Goldschmidt to be the flexible means with which to structure studies of representation. Either axis may provide the lead, while the other axis is derivative or, alternatively, the two may attain equal standing and even describe a matrix, if this is advantageous to the exploration at hand. Interconnections among dimensions and axes are, of course, the outcome of the focus and priorities of every specific study, but the overall scheme attempts to describe a possible paradigm for the study of design representation.

Gabriela Goldschmidt and William L. Porter

PART I

From the Perspective of Architecture

1

Distance and Depth

Penny Yates

Thus the eye which was previously directed towards the left of the church facade, towards the point of entrance, is now violently dragged away towards the right. The movement of the site has changed. The visual magnet is no longer a wall. Now it has become a horizon. And the wall, which previously acted as backdrop to one field of vision, as a perspective transversal, now operates as a side screen to another, as a major orthogonal which directs attention into the emptiness of the far distance but which, by foiling the foreground incident – the three entrails – [*canons à lumière*] all serves to instigate an insupportable tension between the local and the remote. In other words, as the church is approached, the site which had initially seemed so innocent in its behaviour becomes a space rifted and ploughed up into almost unbridgable chasms. ...
It is possible, but it is not probable, that all this is uncontrived.
...

Colin Rowe on Le Corbusier's La Tourette
(Rowe 1976)

In 1961 Colin Rowe wrote about the perceptions of a visitor to Le Corbusier's recently completed monastery, La Tourette. Rather than describing the objective properties of the work, he portrayed it through the perceptions of the peripatetic viewer, revealing that meaning unfolds itself as the relationship between the viewer and the object in view changes. That he wrote this description in the third person suggests that it was not simply a personal experience but one that was available to anyone with some affection for architecture and a willingness to engage the visual experience.

As the "precession of simulacra" observed by Jean Baudrillard (1988) becomes less a cultural criticism and more an accepted reality, the focus of most architectural publications is on the image; indeed, the representations become the reality of the object. Fredric Jameson (1983) has further remarked on the transformation of reality into images with the result that "our entire contemporary social system has little by little begun . . . to live in a perpetual present" represented by these images (ibid., p. 125). Pierre-Alain Croset (1988), writing in his capacity as editor of *Casabella*, expressed a concern that "in these images what disappears is a fundamental dimension of architecture: its temporal experience, which by definition *is not reproducible*" (ibid., p. 201). The result is a "fall of the value of experience [which] clearly manifests itself in the present tendency of architects to under-rate the problems tied to the spatial experience of the building while paying excessive attention to the *external* visual character of the object" (ibid., p. 205). He goes on to say that what is needed and rarely supplied is "the support of a narration – that is, of the only instrument that can evoke what photographs cannot reproduce" (ibid., p. 204). Yet it is rare to find a narration as purposefully based on perceptual experience as Colin Rowe's essay on La Tourette. Rarer still is Rowe's inference that the architect may have contrived a significant and cerebral perceptual experience for the viewer.

Design, including architectural design, is the intellectual conception of a manufactured or constructed object prior to its production. Any discussion about how a designer arrives at this conception generally focuses on the relative importance of process versus goal – a corollary to the scientific dialogue about the relative virtues of the inductive and deductive methods. Some believe that the correct process or method will produce the ideal object; others believe that the designer must somehow know in advance the ideal properties of this object and then seek the means of achieving that ideal. Both the positivist and the idealist positions concern themselves principally with the qualities that inhere in the object, properties that are quantifiable and verifiable. Neither places significant emphasis on the more incommensurable responses of the perceiving subject.

Mathematicians measure the shapes and forms of things in the mind alone and divorced entirely from matter. We [painters], on the other hand, who wish to talk of things that are visible, will express ourselves in cruder terms.

Leon Battista Alberti, On Painting

Representation in the design process is a visual testing of the results of this conceptual process. The necessity of such a test will lie with the designer's goals, but most architects still rely on (and most clients still demand) some visual verification of the concept, especially considering the enormous cost of materializing the concept. But the design representation can be more than a test; it can provide an insight into the designer's process, whether it is a sketch, diagram, model or photograph, or a full-blown rendering, whether it is prior to the execution or a memory. A designer's representation of his or her work, whether it is executed before or after the physical manifestation of that work, can provide an insight into the individual's design process. This chapter will interrogate selected examples of architects' drawings and photographs to identify possible ways in which acknowledgement of subjective perception may inform design thinking. Architectural design is the specific vehicle for this enquiry. Nowhere is it intended to suggest that these approaches to design are mutually exclusive, or that pragmatic and purpose-made properties of objects are dispensable or even secondary to perceptual considerations, but rather that, in the architect's extraordinarily difficult task of integrating a multitude of requirements, each has a potential role.

The distinction between subject and object lies at the core of the Western philosophical tradition. The subject is the being who views the world through individual apparatus. Objects comprise that world. Post-Socratic philosophy conferred on the human the paradoxical properties of being both an observing subject *and* an object in the world of objects. This dualism was exacerbated by the epistemology of the 17th-century philosopher René Descartes, in which the rational mind of the subject is elevated above the physiological senses as a means of comprehending the world. The Cartesian *cogito* is thus the subject, or self, relegated to an existence within its own intellectually constituted world. Severed from the world of sense perceptions, the *cogito* is disembodied – a kind of universal Being cerebrally observing the world's objects in a rationalized space – uniform and infinite.

Two very different strains of thought have characterized the ever-increasing distance between the subject and the objective world that resulted from the Cartesian mind–body dualism. The first resulted in the technological scrutiny undertaken by the 19th-century Positivists who exploited their intellectual distance from the objective world. By dissecting and classifying the objective world into constituent parts, they sought to discover ever more precise truths about the workings of nature and the universe. The other strain of thought considered the objective world to be extraneous to the better workings of the subjective mind. Nineteenth-century Romanticism assumed that creative endeavours were stimulated by individual intuition and invention. Concomitantly, a new concept emerged in which aesthetic beauty was no longer inherent in the object but determined by the eye of the beholder. The lingering legacy of this 19th-century distinction, Positivism and Romanticism, has continued into the 20th century, and this confrontation, broadly speaking, characterizes the popularized notions of the differences between science and art. It is perhaps in the design fields where the conflicts inherent in this distinction have their greatest impact, as the designer sees himself/herself at one moment as a scientist, at the next as an artist.

Fundamental to much of 20th-century philosophy is the disavowal of the Cartesian mind–body dualism. It is held by many to be responsible for much of what is objectionable about our current technocratic state. As Martin Jay (1994) has documented in his book, *Downcast Eyes: The Denigration of Vision in Twentieth-Century French Thought*, the fear of residual Cartesianism has engendered a nearly insurmountable distrust of vision because of its association with the *cogito* – the disembodied subject. Likewise, the Romantic conception of subjectivity is distrusted for its visual relativism and eccentricity.[1] The architect's design process at its best is a conceptual act. However, when the conceptual is made material, the visible environment is significantly altered. Although there are numerous other conditions influencing the act of design, it is foolish, if not dangerous, to disregard the visual results. Many architects would agree with Steven Holl (1980) that "Architecture . . . [is an] essentially wordless art – that is, . . . [an] art that lies just beyond the reach of words" (ibid., p. 71). Indeed, the architect must rely on visual representations to present or reinforce those aspects of the design that are impossible to convey with words alone. The issue, then, is not to circumvent the visual because of its metaphysical and epistemological associations, but rather to discover how we might redress the Cartesian disembodiment of the subject and restore interest in the subjective experience of the visible world.

At last I can see as God sees!

Leon Battista Alberti

In 1925 Erwin Panofsky postulated in his short but influential text, *Perspective as Symbolic Form*, that perspective, or any other pictorial "treatment of space," is not simply an abstract tool for two-dimensional representation, but a symbol[2] for the contemporaneous metaphysical understanding of the world. What is most provocative about Panofsky's essay, however, is the observation of an historical reconciliation between the subject and the object. According to Panofsky, the result of the mathematical "systemizing of space" in Renaissance pictorial representation was an "objectification of the subjective" (1991, p. 66) view of the world.[3] From Filippo Brunelleschi's first public demonstrations in Florence, perspective technique was intentionally mimetic. The philosophical question of whether the pictorial representation that awed the *quattrocento* onlookers was a representation of Brunelleschi's subjective experience or a reproduction of a purely objective condition is a diachronic one. In the 15th century there was man and there was God; the contemporary distinction between subject and object did not exist in the epistemology of the time. This vexing dualism required Descartes to make it manifest and Immanuel Kant to articulate it for the metaphysics of the modern world.[4]

A recent book by Alberto Pérez-Gómez and Louise Pelletier, *Architectural Representation and the Perspective Hinge* (1997), is synchronically premised on the belief that the tools of architectural representation are never neutral, but are ineluctably linked to the metaphysical and epistemological beliefs of a particular time and place. These tools, and therefore these beliefs, "underlie the conception and realization of architecture" (ibid., p. 3). Their book documents a history of representation in which there are numerous examples of architects and artists who, in their theories and their representations, have consciously considered the subjective viewer. Pérez-Gómez and Pelletier share contemporary philosophy's critical view of Cartesian epistemology, noting that our tools of representation have been used increasingly as instruments to document the objective world as intellectually constructed by the detached, "disembodied" observer. The subjective point of view, they believe, has been purged from representation, while techniques such as isometry, axonometry, and, currently, computer drafting proliferate (Figures 1.1 and 1.2). These methods of representation rely on a spatial construct of absolute measurement in which the world's objects are contained and located without regard for the viewer's station point or the surrounding context.

The tragedy of our times is that measures have everywhere become abstract or arbitrary; they should be made flesh, the living expression of our universe, ours, the universe of men, the only one conceivable to our intelligence.

Le Corbusier, The Modulor

In response to the hegemony of this technological mode of vision and representation, Pérez-Gómez and Pelletier appeal for the recovery of meaning in architecture that can be discovered in human experience. They acknowledge the influence of the phenomenal ontology of the late French philosopher Maurice Merleau-Ponty, which endeavoured to revitalize the visuality of a reconciled mind and body. Merleau-Ponty certainly shared with his contemporaries the suspicion of the Cartesian *cogito* as a means to knowledge: "Intellectualism and empiricism do not give us any account of the human experience of the world; they tell us what God might think about it" (1996,

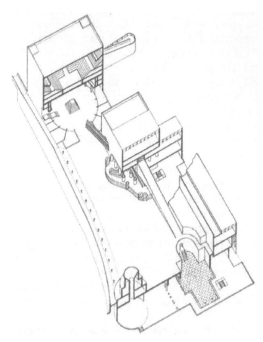

Figure 1.1 Office of James Stirling, The Wallraf Richartz Museum Competition, Cologne, 1975. James Stirling/Michael Wilford Archive © Collection Centre Canadien d'Architecture/Canadian Centre for Architecture, Montréal.

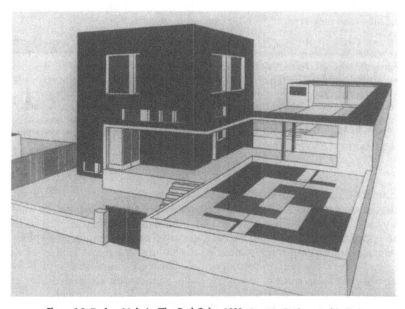

Figure 1.2 Farkas Molnár, The Red Cube, 1923. Courtesy Bauhaus-Archiv Berlin.

p. 255). Seeking to reconcile the mind and the body, he offered an alternative epistemology: "My body is the fabric into which all objects are woven, and it is, at least in relation to the perceived world, the general instrument of my 'comprehension'" (ibid., p. 235). To become more familiar with this modern subject, whose body inhabits the objective world ("the world is made of the same stuff as the body" (Merleau-Ponty 1954, p. 163)), who is the "instrument of [my] comprehension" in the nearly forsaken ground of the perceived world, the words of Merleau-Ponty will be our guide.

If, as Pérez-Gómez and Pelletier have posited, the embodied observer has been systematically eliminated from architectural representations, by what means might the presence of the subject in a representation be revealed? To begin to differentiate between the objective and the subjective in architectural representation, the idea of "distance" will be distinguished from that of "depth." The term "distance" will be used as an agent of objectivity capable of describing the measurable properties of an object and its location. The term "depth" will be used to describe the experience of the embodied subject locating oneself in a world of objects in relationship to these objects. It suggests that architectural form can be conceived and described in perceptual terms, in relationship to the viewer. In the words of Merleau-Ponty, depth "is not impressed on the object itself"; rather, it "announces a certain indissoluble link between things and myself by which I am placed in front of them" (1996, p. 256).

In their seminal essay "Transparency: Literal and Phenomenal" (1963), Colin Rowe and Robert Slutzky discuss the perception of depth in architecture. The authors' point of departure is Siegfried Giedion's *Space, Time and Architecture* (1967) in which he proposed an epochal commonality between early Cubist vision and transparency in architecture – the latter being the rather unambiguous ability to see from one location through a transparent membrane to another. Rowe and Slutzky (1976) compare this definition to Gyorgy Kepes' description (Kepes 1944, p. 77) of two-dimensional spatial ambiguity that they refer to as phenomenal transparency:

If one sees two or more figures overlapping one another, and each of them claims for itself the common overlapped part, then one is confronted with a contradiction of spatial dimensions. To resolve this contradiction one must assume the presence of a new optical quality. The figures are endowed with transparency; that is, they are able to interpenetrate without an optical destruction of each other. Transparency, however, implies more than an optical characteristic; it implies a broader spatial order. Transparency means a simultaneous perception of different spatial locations. Space not only recedes but fluctuates in a continuous activity. The position of the transparent figures has equivocal meaning as one sees each figure now as the closer, now as the further one.

(ibid., pp. 160–161)

Phenomenal transparency shares with its literal cousin the "simultaneous perception of different spatial locations" but, instead of a seeing through to what is beyond, there are both equivocal and terminated readings of depth. In architecture the ambiguous and fluctuating spatial order is produced by physical surfaces in varying three-dimensional relationships *to the viewer* – i.e., Merleau-Ponty's "certain indissoluble link between things and myself by which I am placed in front of them." Actual or numerically defined, distances between these surfaces are irrelevant to the perception of their locations and relationships. Perhaps the most important distinction is that Giedion's Cubist analogy suggests that literal transparency is a property of the architectural artefact that can be dissected and classified along with other observable properties. Phenomenal transparency, on the other hand, is not a property of the objects constituted by these surfaces; the presence of a viewing subject is required for the effects of equivocal depth to be perceived.

The visible deeply objects to our habitual objectification.
David Michael Levin, The Opening of Vision

In 1968 the Swiss architect Bernhard Hoesli, who had been on the faculty at the University of Texas at Austin with Colin Rowe in the 1950s, republished the Rowe and Slutzky essay in German. In sincere appreciation of their concept of transparency "as a tool for study [which] makes understanding and evaluation possible," Hoesli wrote his own accompanying essay, copiously illustrated, in which he suggested that [phenomenal transparency] was an "employable, operative means enabling the intellectual ordering of form during the design process, as well as its graphic representation" (Hoesli 1997, p. 60). However, his diagrammatic restatement of their ideas seems to have missed their point. He illustrated the layering of planes in one of Le Corbusier's Purist paintings from a lateral, or flanking, as opposed to a frontal point of view (Figures 1.3a and b).

A clearly ordered Cartesian spatial matrix regulates the distance from one plane to the next, but all allusion to perceptual depth has been eliminated from this disembodied view of the painting. Hoesli then applied the same diagrammatic formula to Le Corbusier's Villa Stein, erecting the receding planes with distances between determined, no doubt, by the numeric fact of the plan (Figures 1.4a and b).

While devising a visualization of overlapping planes, Hoesli succumbed to what Merleau-Ponty called the "traditional view" – the habit of objectification – "in which experience of depth . . . consists in interpreting certain given

Figure 1.3a Le Corbusier, *Nature morte à la pile d'assiette et au livre*, 1920. Peinture FLC 305 © 2003 Artists Rights Society (ARS)/ADAAGP, Paris/FLC.

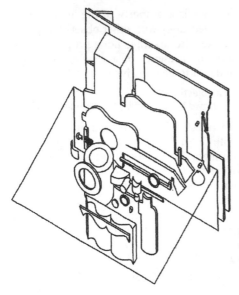

Figure 1.3b Diagram by Bernhard Hoesli of Le Corbusier's *Nature morte à la pile d'assiette et au livre.* Dessins de B. Hoesli. Courtesy FLC. © 2003 Artists Rights Society (ARS)/ADAAGP, Paris/FLC.

Figure 1.4a Le Corbusier. Paris: Villa Stein 1927. © 2003 Artists Rights Society (ARS)/ADAAGP, Paris/FLC. Photograph by F.R. Yerbury. © F.R. Yerbury/Architectural Association.

facts – the convergence of the eyes, the apparent size of the image, for example – by placing them in the context of objective relations which explain them" (1996, p. 257). Elsewhere, Merleau-Ponty said: "In order to treat depth as breadth viewed in profile, in order to arrive at a uniform space, the subject must leave his place, abandon his point of view on the world, and think himself into a sort of ubiquity" (ibid., p. 255). We know, of course, that "readings" of depth are facilitated by the fact that some planes (and some objects)

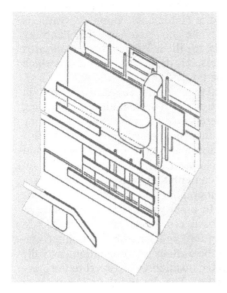

Figure 1.4b Diagram by Bernhard Hoesli of Le Corbusier's Villa Stein. Dessins de B. Hoesli. Courtesy FLC. © 2003 Artists Rights Society (ARS)/ADAAGP, Paris/FLC.

are more distant from the viewer than others. But the perceptual qualities of that reading that were so critical to Rowe and Slutzky's analysis are reduced to hypothetical relationships and are no longer the concern of Hoesli's analysis. Thus he alters the wording of Kepes' definition of transparency to say:

Transparency arises wherever there are locations in space which can be assigned two or more systems of reference – when the classification is undefined and the choice between one classification possibility or another remains open.

(Hoesli 1997, p. 61)

Kepes, Rowe and Slutzky do not ask us to classify the planes or distantiate their locations; they ask us to see them in relation to ourselves and to each other, to consider their perceptual interaction, to understand their depth. As Merleau-Ponty ruminated in his working notes for *The Visible and the Invisible*: "Depth . . . is pre-eminently the dimension of the hidden . . . of the simultaneous. Depth is the means the things have to remain distinct, to remain things. . . . Whereas by virtue of depth they coexist in degrees of proximity, they slip into one another and integrate themselves. It is hence because of depth that the things have a flesh. . . ." (Merleau-Ponty 1968, p. 219).

Like so many other Corbusian elements they are obedient to the exigencies of the eye rather than those of the work, to the needs of the conceiving subject rather than the perceived object.

Colin Rowe on Le Corbusier's La Tourette

If depth is not a property of an object but a subjective perception, is it possible to identify examples of representations that are concerned with revealing perceptual depth rather than objective distance? On the surface it would seem that the perspective view, the view that Panofsky identified as "the objectification of the subjective," would be just such a representation. However, among the contemporary thinkers who have sought to disinherit the *cogito* from our world-view, many, including Merleau-Ponty with his reference to "the traditional view," have condemned perspective representation as a mathematically constructed view of the world, a Platonic deception, a "synthesis of experience into mental abstraction."[5] The subjective view that Panofsky assumed to be present in perspective has, for the most part, been eradicated by our habit of objectification. Nevertheless, it is the perspective view in which the perceiving subject can most easily be implied.

In the absence of texts that make reference to the subjective viewer, all evidence must be sought in the artefact of the representation. In perusing published sketches for traces of the presence of the subjective view in the generative stages of design, it seems that there are exceedingly few examples.[6] The following examples are, for the most part, expository drawings and photographs that document the design object, whether completed or in development. In searching for such documentation, it is surprising to see how frequently such drawings and photographs, though devised perspectivally, do not admit a corporeal viewer. In other words, the station point of the perspective view is one that is impossible or irrelevant with respect to the experience of an actual body coming into contact with the architectural artefact. One may notice, for example, how many representations actually place the station point in a slightly elevated position with respect to the gravitational ground plane (Figures 1.5 and 1.6). The eye and the mind are thus disembodied. The scene is an abstraction of the mind's eye; the corporeal viewer is absent.[7] If depth *is* visible in such representations, the question is, *to whom?* The point of view in these images is devised for the observer of the simulacrum.

As Martin Jay has remarked, Merleau-Ponty attempted to save perspective from the fate of intellectual assimilation with his concept of nontranscendental perspective that "reunited [humans] with the objective world" (1994, pp. 303–304). He rejected the notion that all such images are enframed and therefore privileged when he said: "The fixed point is not made by intelligence: the looked-at object in which I anchor myself will always seem fixed, and I cannot take this meaning away from it except by looking elsewhere." (Merleau-Ponty 1964, p. 52). Thus another avenue to the discovery

Figure 1.5 Louis Kahn, project for the Philadelphia Midtown Civic Center Development. Courtesy the Louis I. Kahn Collection, The University of Pennsylvania and the Pennsylvania Historial and Museum Commission.

Figure 1.6 Louis I. Kahn, Mikveh Israel Congregation, Philadelphia, 1963. Digital Image © The Museum of Modern Art/Licensed by SCALA/Art Resource, NY.

of traces of the subjective viewer in architectural representations is to seek effects that occur only in the presence of a hypothetical viewer. Two-dimensional representation can never simulate experience. Nevertheless, we can perhaps derive from these examples something that is absent in so many others.

As may already be surmised from the above examples of representations that lack an implied, incarnated viewer, one condition that can implicate such a viewer is the presence of a ground plane that makes the subject's location comprehensible. One indication is isocephaly,[8] or the horizontal alignment (more or less) of the viewer's horizon line and the heads (and therefore eyes) of other humans, indicating that they share a common ground plane. The proto-perspectivalists of the 14th and 15th centuries, such as Giotto, or the painters of the Sienese school, for example, intuitively struggled with this observed phenomenon as they increasingly tried to depict the ideal using the characteristics of the physical world. Karl Freidrich Schinkel frequently composed his perspective views in this way. A typical example is the engraving of his design for Neuen Schauspielhaus (Figure 1.7). It can be seen from this image that the exploitation of this technique depends on a flat ground plane. This image was chosen in particular to contrast the various figures that occupy the two-storey flight of steps and the portico to those that occupy the same ground plane as the viewer/artist. Architectural elements such as doors and windows, which indicate human scale, can also disclose that the viewer occupies the represented ground plane.

Readings of depth, which disclose "the link between the subject and space" (Merleau-Ponty 1996, p. 267), however, are suppressed because of the distance of the viewing subject from the object, the lack of foreground, and

Figure 1.7 Karl Freidrich Schinkel, Neuen Schauspielhaus. Ferdinand Berger, delineator.

the emptiness of middle ground. The human figures in this representation share the background with the edifice, which miniaturizes them in its presence. Nevertheless, the viewer, while forced to remain disjunct from the objects represented, must be struck by the same imposing scale that the depicted figures must experience. Schinkel was certainly more influenced by the emerging ideas of 19th-century Romantic subjectivity, including the contemporary embracing of the sublime, than by any notion of subject–object reconciliation. Isocephaly fixes the viewer's position in the vertical dimension; one's feet are essentially on the ground. It cannot, however, determine the subject's lateral position relative to the object, and there is nothing about Schinkel's engraving that suggests that the station point was determined by a subjective point of view. Indeed, the three-quarter view of the Neuen Schauspielhaus is devised to best "show" the object; its monumental façade appears to "face" some other grand gesture in the objective world, consciously turning itself away from the viewer.

Generally, it would be agreed that, in the first half of the 20th century, architectural classicism finally gave way to modernism, but in recent decades a number of architectural critics have concentrated as much on the continuity as the breach. For example, the latent classicism in the work by Mies van der Rohe, one of the great modernists, has been demonstrated on many occasions. After visiting Mies' reconstructed German Pavilion in Barcelona, Robin Evans observed that "the most striking properties of the pavilion have to do with the *perception* of light and depth" (1997, p. 256) rather than its "sublime rationality" (ibid., p. 244) or the "*transcendent logic of its determining grid*" (ibid., p. 246; emphasis added). Perhaps the most recognizable feature of most of Mies' smaller buildings is the emphasis on the horizontal plane. It is not unusual to see photographs where the sensation of depth is captured by the receding lines of "floating" ceiling planes above and the grid of the floor pavers below. These planes are typically placed at more or less equal distances from the eye level of the average viewer, resulting in a top and bottom symmetry about the horizon line (Figure 1.8). One cannot deny this fact of the design, but was it intended *for the purpose* of the viewer's perception?

According to Evans, the more subtle and, he suggests, probably more deliberate effect at the Barcelona Pavilion is the purely visual *restoration* of

Figure 1.8 Ludwig Mies van der Rohe. The Bacardi Administration Building, Mexico City. Photography by Werner Blaser.

classical bilateral symmetry in this asymmetrically planned composition. By virtue of reflection, the mirrored horizontal and vertical planes align themselves with their sources to create new, widened and elongated planes (Figure 1.9). The line of symmetry is no longer an abstract centreline but a physical plane captured between the horizontals which appear to bypass or to cut through the vertical planes. The depth that is apparent in the non-reflected view is visually compressed because of the change in proportions of length to width of the image.

Figure 1.9 Ludwig Mies van der Rohe, The Barcelona Pavilion (reconstruction), 1986. Photograph by Robin Evans. Courtesy Janet Evans.

In a statement similar to Colin Rowe's on Le Corbusier, Evans claims for Mies: "if [he] adhered to any logic, it was the logic of appearance. His buildings aim at effect. Effect is paramount"(ibid., p. 247). Unlike the placement of horizontal planes symmetrically above the typical viewer's horizon line, the visual effect of reflective symmetry is not only devised for the viewer; it depends on the viewer's presence for its existence. But the fact of its presence cannot answer the question regarding Mies' design process. Is this effect simply a fortuitous perceptual occurrence or was it an intentional component of the design?[9]

The perception of depth is not available in the Cartesian conception of space as uniform and infinite. An orthographic drawing (e.g., a plan, section, or elevation), or an isometric drawing that is constructed of real proportional dimensions and relative coordinates cannot provide the designer any insight into the qualities of appearance to the subjective viewer. Of course, a designer's experience permits valid speculation about the visual results of an abstract representation. But if it is of any import to the designer to test for such visual qualities as depth prior to the materialization of a design, a different kind of representation will be made.

The invention (or recognition) of modern perspective techniques for constructing a painting or a drawing, first articulated by Leon Alberti's codification of the experiments of Brunelleschi and others in 15th-century Italy,[10] relied essentially on three fundamental concepts: first, that lines that are parallel appear to converge; second, that these lines converge to a single point, which Alberti called the centric point; and third, that the apparent decrease in distance between equidistant transverse lines could be determined by geometric method. To be sure, Alberti himself understood the illusory nature of this form of representation when he stated: "No learned person will deny that no objects in a painting can appear like real objects. . . ." (1991, p. 56). But this mathematical artifice immediately and overwhelmingly became the accepted convention for artistic accuracy in portraying the earthly world. The clearest evidence of the acceptance of this convention was the soon-to-be ingrained ability of painter and viewer alike to apprehend converging lines as parallel and transverse lines of diminishing distance as in fact equidistant. The diagonal line became virtually synonymous with depth.

The numerous treatises on perspective construction that followed its early practice were almost without exception mathematically based until the invention of photography fixed the perspectival image on a flat surface without the aid of the human optical organ. This development caused the discourse to shift to an examination of whether perspective construction is the nearequivalent of the optical conditions of sight or whether it is merely a convention of Western representation that has been so culturally ingrained that it prefigures the way we see. As previously discussed, in the 20th century the perspective view has been frequently discredited by those who consider it a purely intellectual and abstract construction of the objective world. As such it is infinite,[11] totalizing, and, most importantly, the privileged view of the contemplative gaze of a disembodied viewer. Further discussion of this contentious issue need not be pursued here since there are numerous other texts to consult.[12] It is true, however, that perspective is not intrinsic to objects. Even though this subjective vision can be simulated mathematically, it requires at least a conceptual subject in order to exist at all. And surely it is conceivable that the perspective view can have sources other than the contemplative gaze of the *cogito*. The perspective view is available as well to the carnate subject who sees not infinity, but depth – finite and modulated. The world is not spread out before the embodied viewer; instead, in the perspective view, much of what is available to be seen can be only partially seen.

The ground that I do not see continues nonetheless to be present beneath the figure which partially hides it.

Maurice Merleau-Ponty, Sense and Non-sense

This is not to say that the viewer consciously reconstitutes the objective totality of all that is perceived. An example is the classic technique of spatial connection in architecture known as enfilade, in which centrelines of doorways or openings to a series of spaces are aligned (Figure 1.10). This simple phenomenon, associated with one-point perspective, has been popularized by frequent recording in paintings and photographs; its perception does not require an intellectual comprehension of perspective techniques. Yet these depictions demonstrate that depth is visible even when the physical distances are compressed into two dimensions.

The trained eye can actually discern this phenomenon in an orthographic drawing (Figure 1.11). If, for example, a plan of a building shows a series of adjacent spaces, and the openings between these spaces are aligned and more or less equidistant, the experienced architect will mentally translate this plan notation into the compressed image of the resulting perspectival view. The plan provides a description of the whole, if only in two dimensions.

But a description of this visible phenomenon that produces depth is quite different. Multiple frontal planes phenomenally recede from foreground to middle ground and ultimately to background. While most of the surfaces of

Figure 1.10 Carlo Scarpa, Castelvecchio, Verona. Photograph © Antonio Martinelli, 1985.

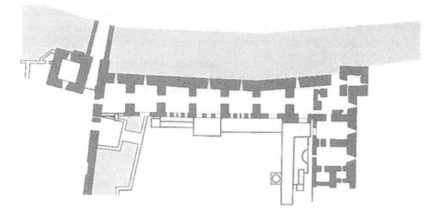

Figure 1.11 Carlo Scarpa, Castelvecchio, plan view.

these receding planes are concealed by the plane in front, the viewer never-theless apprehends the series of spaces that are clearly laid out in the plan view. Further, most of the converging diagonals that usually convey depth are also concealed from view and must be inferred, consciously or unconsciously, by the painter, photographer, or viewer. In short, much of what is available in the objective world is omitted from this subjective view of it.

All perspective views depend on a fixed station point, or position, of the viewer. There exist a limited number of station points that allow the viewer to perceive architectural spaces in enfilade. One must assume a point of view near one end of the series, and the centre point, or vanishing point, must fall within all the openings to the space. In other words, to record this phenom-enon, the viewer must place himself or herself in a position of "registration" with the architecture. The result of this registration is an abstracted visual axis that runs from the viewer, through all the openings, and is terminated in a frontal plane that is at the greatest distance from the viewer. If the viewer moves a few degrees laterally in either direction, the "registration" will be lost and the phenomenon will no longer be visible. The casual viewer, who is unlikely to be cognizant of the abstracted "visual axis," will certainly recog-nize it as the axis of circulation. Thus the enfilade view is not a simple phe-nomenon for the contemplative gaze; it compels the viewer to travel through the perceived space, to travel along the axis to its end. So, too, is this experi-ence reversible; the viewer will experience the same phenomenon on return.

One-point perspective is frequently used to represent architecture, espe-cially architecture that is more or less bilaterally symmetrical. Here an equiv-alency arises between the line of symmetry, which is a geometric abstraction, and the visual axis between the viewer and the vanishing point. What then determines the choice of the viewer's station point – the geometry of the object or the desire of the viewer to place himself or herself in "registration" with that geometry? Does the effect inhere in the object or must the subject actively engage the object to discover this registration?

An example of the ramifications of enfilade as a geometric technique is Louis Kahn's Salk Institute in La Jolla, California. Kahn's lectures and writings, which reveal him to be one of the great humanist architects of the 20th century, have inspired legions of architects throughout the world. His legacy of drawings is remarkable for depicting the realm of human activity within the spaces and places he designed. Yet the numerous perspective views of his projects are typically composed without a corporeal viewer in mind (Figures 1.5 and 1.6). The viewer's station point for the perspective view is almost always located indeterminately. Kahn was trained as an architect in the United States, but he was most influenced by professors and practitioners who were trained in Paris at l'Ecole des Beaux Arts, where emphasis was placed on the development of the plan. Throughout his career his own plans were informed by axial strategies for connecting geometric volumes; he was undoubtedly aware of the resulting visual effect of enfilade. At Salk the technique explodes into an architecture of numerous local symmetries and a multiplicity of spaces in enfilade (Figures 1.12 and 1.13), treating the viewer to one opportunity after another to find a "registration" with the architecture. Here is a place where interaction with the architecture inspires subjective self-awareness. However, there is little positive evidence that Kahn himself thought consciously about achieving these visual effects.

Man looks at the creation of architecture with his eyes, which are 5 feet 6 inches from the ground. One can only consider aims which the eye can appreciate and intentions which take into account architectural elements.

Le Corbusier, Towards a New Architecture

There is every reason to believe that Le Corbusier did think explicitly about the visual effects of his architecture and the engagement of the subject. In addition to statements such as the one quoted above, his photographs employ several techniques in which one can discern the presence of the viewer. It should be clarified here that the photographs discussed in the following paragraphs (Figures 1.14 and 1.15) were created by Le Corbusier to promote his design work. They were composed with deliberation and frequently cropped

Figure 1.12 Louis Kahn, The Salk Institute, La Jolla, California. Photograph by Penny Yates.

Figure 1.13 Louis Kahn, The Salk Institute, La Jolla, California. Photograph by Penny Yates.

for publication. The observed effects are almost certainly intentional. He frequently composed his photographs using one-point perspective from the point of view of an average height person occupying the represented ground plane. However, he freed this compositional technique from its association with bilateral symmetry by identifying the *perceptual* condition that establishes depth – the visual axis[13] – *the abstract line that connects the subject's eye to a focal point*. If this connection is of sufficient perceptual strength, it no longer requires *two* opposing but symmetrical edges to establish its presence. This perceptual axis replaces the abstract and objective mathematical axis of symmetry; the architecture no longer needs to be equivalent on both sides. Nor does this axis necessarily double as a circulation axis. In Le Corbusier's photographs one frequently finds objects located along or near the visual axis to deny the traditional equation between the path of the eye and the path of the feet. These departures from tradition were appropriate to Le Corbusier's quest for a new architecture for the modern world. They were also in keeping with his explorations of spatial ambiguity in his post-Cubist *purisme* paintings of the 1920s and 1930s.

The visual axis connects a viewer and a focal point. Since the viewer's location is not necessarily dependent on the particular geometry of the unified architectural object, as in the view of spaces in enfilade, the possible station points will instead be determined by the viewer's discovery of, or registration with, a focal point. The focal point is unlike Alberti's abstract centric point, the purpose of which was to construct a mathematically correct representation. Rather, the focal point is like the "visual magnet" in Colin Rowe's description of La Tourette quoted at the beginning of the chapter, or Merleau-Ponty's fixed point of "any looked-at object in which I anchor myself." The focal point and the vanishing point are virtually synonymous in these photographs, but they are not equal. The focal point is primary, contrived to locate the subject's focus in the image; its mathematically precise location is irrelevant and it may be adjusted for purely compositional purposes. The vanishing point is determined by, and after, the focal point. In the photographs the lines converging towards the vanishing point reinforce the presence of a focal point but are not specific to it. A focal point may be apprehended at varying proximities to the viewer; the vanishing point is, mathematically speaking, at an infinite distance. The focal point generally resides in a plane that is frontal to the viewer, the axis then being perpendicular to this plane. This plane terminates the reading of depth so that the viewer may apprehend shallow or deep space.

In these photographs the visual axis is cultivated as an axis of asymmetry. Architectural elements assemble themselves about the visual axis without regard for symmetrical arrangement. Visually, the receding frontal planes, which disclose to the eye only a partial view of the space beyond, can be juxtaposed to the lateral planes that are parallel to the visual axis, but are represented perspectively as converging diagonals. On one side of the visual axis, the reading of depth is more dynamic, as the eye races from the foreground to the background. On the other side, it is more modulated by the spatial transversals of the coulisses between the frontal planes. Both conditions begin to beckon the viewer into the scene which is not scenographic but the space of potential occupation.

Figuratively, with the introduction of asymmetry, the possibilities abound to create new meanings using binary commentary. For example, at the entrance to the Villa Savoye (Figure 1.14) the camera's/viewer's eye is fixed on a focal point in the distant landscape (ambiguous in its definition, but unmistakable in its presence). The visual axis between the viewer and the focal point is roughly equivalent to the centreline between a row of columns on the right and a transparent wall on the left. Narratively, it seems clear that the viewer is arriving and not leaving since the entry door is forward in the route; however, there is the insistence of the focal point in the landscape, producing a tension between inside and out, between arriving and continuing, between culture and nature.

Similarly, in a view from the roof garden towards the ramp, the visual axis virtually divides the photographic composition into two halves (Figure 1.15).

Figure 1.14 Le Corbusier. Poissy: Villa Savoye 1929 © 2003 Artists Rights Society (ARS)/ADAAGP, Paris/ FLC L2(17) 23.

Figure 1.15 Le Corbusier. Poissy: Villa Savoye 1929 © 2003 Artists Rights Society (ARS)/ADAAGP, Paris/ FLC L2(17) 49.

Clearly, the focal point is the dark door in the white wall at the centre of the image (one almost assumes that the perspectival vanishing point must be in the centre of the door, but in fact it is very much closer to the left edge). Reinforced by the severity of the diagonal convergence of the lateral wall between living-room and roof garden, the connection between the viewer and the focal point is riveting. The axial cleavage creates an opposition between inside and outside, dark and light, spatially determinate and spatially indeterminate. The visual axis, which impels the viewer forward, is countered with a different kind of perceptual diagonal denoting the more objective, upward movement of the ramp. The ramp and the spiral stairs are captured in the coulisses between the layered, frontal planes. The expressed depth on the left side competes with the implied depth of the frontal planes on the right. Both vie for the viewer's attention and potential movement; horizontal is contrasted with vertical.

This fascination with the bilaterally asymmetrical led Le Corbusier to more complex and less traditional photographic constructions. Thomas Schumacher (1987) has demonstrated that a large number of photographs in Le Corbusier's *Oeuvre Complète* borrow the compositional device used by Piero della Francesca in his painting *The Flagellation* in which a single image is turned into a virtual diptych (ibid., p. 41).

In Piero's painting a column divides the composition spatially into a right side and left side (Figure 1.16). The vanishing point[14] of the composition occurs within the left side of the painting and, while it does serve to establish a space of relatively great depth, it does not function as a focal point as described above. The eye is not drawn to the vanishing point; indeed, the vanishing point can be ascertained only by analysis of the converging lines. Here there are competing focal points, one on each side of the painting, created by two groupings of human figures, each of which *en masse* creates a distinct frontal plane and a focal point. The tension between these two focal points is exacerbated by the positioning of one group at the extreme foreground and the other in the background. This is not the place to reference the many

Figure 1.16 Piero della Francesca, *Flagellation of Christ*, Galleria Nazionale delle Marche, Urbino. SCALA/Art Resource, NY.

interpretations of Piero's artistic device, especially given the subject matter. Rather, it will suffice to say that Piero's methods were almost certainly more in the service of the narrative than in the composition. His entire *oeuvre* consists of tableaux that are temporally transcendent; spatial ambiguity rather than spatial logic reinforces this narrative intention.

In his essay Schumacher demonstrates that many of Le Corbusier's images are composed "as an overt and conscious act" (ibid., p. 41) using this bilateral juxtaposition of two focal points, one in deep space and one in shallow space. He further analogizes the composition of these photographs to the post-Cubist Purist paintings of Le Corbusier and his colleague Amédée Ozenfant in which orthographically derived compositions are rendered spatially ambiguous, as planes in closer proximity overlap more distant planes, creating spaces of ambiguous dimensions, as deep space continues invisibly behind shallow space, and as figure and ground merge and fluctuate. In Le Corbusier's photographs a similar, ambiguous merging of foreground and background causes a "spatial collapse."

But there is another dynamic at work in these photographs. As a mathematical construction, one-point perspective admits only one vanishing point. This was insufficient for the spatial complexity of Le Corbusier's modernist architectural intentions, as well as his artistic interests, in which he (like the Cubists before him) pursued spatial ambiguity, dispersal of focus, and compression of depth. As in the previously discussed photographs, Le Corbusier relinquished the power of the mathematically determined vanishing point to the greater power of the subject's focal point. Piero had done virtually the same thing. As an early perspectivalist, Piero was not beholden to the orthodoxy of the centre point; it was only one among many methods then being used to achieve pictorial depth.[15] In the painting there is a positive spatial and perhaps temporal tension between the two focal points. On the other hand, in the photographs that juxtapose shallow space and deep space, the camera lens will not yield its complicity in the reification of the perspectival vanishing point. To undermine its power, to press the spatial ambiguity and tension between competing focal points, Le Corbusier used several tricks in his photographic constructions. One was to conceal the vanishing point by locating it behind the frontal plane (Figure 1.17). Another was to emphasize frontal planes and limit the presence of the converging diagonals of the lateral planes, as he had done so often in his Purist paintings (Figure 1.18). Yet another was to place the vanishing point in that very element – a tree, a column, a window mullion – that, like Piero's column, divides the image into two parts – deep space and shallow space. The eye must move away from the insidious grasp of the vanishing point to the right side or the left side of the image.

Figure 1.17 Le Corbusier. Le Celle Saint Cloud: Maison de week-end 1934 © 2003 Artists Rights Society (ARS)/ADAAGP, Paris/FLC L1(6) 146.

Figure 1.18 Le Corbusier. Chandigarh: View of the Palace of the Assembly from the High Court 1950–65 © 2003 Artists Rights Society (ARS)/ADAAGP, Paris/FLC L3(10) 209.

My mobile body makes a difference in the visible world, being part of it; this is why I can steer it through the visible . . . Vision is attached to movement.

Maurice Merleau-Ponty, "Eye and Mind"

Unlike Piero's tableau, in which time is symbolized but is discontinuous, Le Corbusier's use of the deep space/shallow space juxtaposition is more dynamic, implying the continuous temporality of the viewer's movement. If one can discern the presence of the subjective viewer, then it is the shallow space that the viewer occupies and it is the deep space that he or she can or will occupy in the future. As in the enfilade view, the expressed or implied receding diagonals of the perspective view compel the corporeal subject to enter the scene rather than remain estranged from it in a distanced, contemplative gaze. Unlike the enfilade, these photographic constructions depict a "here" and a "there"; the space in between is the space of the subject's movement. For the Cartesian subject, unseen objects no longer exist except in memory and judgement. In this view the "here" will continue to exist even when the subject moves towards the "there." And the "here" will be reversible. It will become the "there" when the subject arrives.

The enigma is that my body simultaneously sees and is seen. That which looks at all things can also look at itself and recognize, in what it sees, the "other side" of its power of looking. It sees itself seeing. . . . It is not a self through transparence, like thought, which only thinks its object by assimilating it, by constituting it, by transforming it into thought.

Maurice Merleau-Ponty, "Eye and Mind"

This latter effect might be termed "co-location," the viewer's ability to *conceive* of oneself as occupying a space other than that which is currently occupied. In other words, the representation not only includes the trace of the viewer, but insinuates the possibility of the viewer's movement and the dimension of time. The seer who can see itself can also conceptualize being on the "other side," seeing itself where it is. In other words, one can imagine occupying another space and seeing from it the space that one currently occupies.

This . . . house will be rather like an architectural promenade. You enter: the architectural spectacle at once offers itself to the eye. You follow an itinerary and the perspectives develop with great variety, developing a play of light on the walls or making pools of shadows. Large windows open up views of the exterior where the architectural unity is reasserted.

Le Corbusier, Oeuvre Complète, vol. 1

Many architects have acknowledged the influence of Le Corbusier's idea of the architectural promenade – an itinerary or route through the work of architecture – by referring to it in descriptions of their own work. Less influential are Le Corbusier's actual words about the unfolding of perspectives, suggesting the consistent visual engagement of the viewing subject. In Le Corbusier's description of Maison La Roche-Jeanneret, written for the initial volume of his *Oeuvre Complète*, there is a deliberate obfuscation of the distinction between objective properties and subjective response. Are the "play of light" and the "pools of shadow" objective facts? Or are they dependent on the observing, itinerant subject for their existence? Some years later at the Villa Stein-de Monzie, the architectural promenade became a purposeful assemblage of a series of images, each with bilateral focal points, one in the foreground to locate the viewer and one in the background to attract the viewer along the next segment of the promenade. The views are devised to inspire the movement of the subject through the territory of the architecture. But, rather like a motion picture story-board or Zeno's paradox of the arrow in flight, the sequence is conceived as a series of fixed views, rendering the viewer's motion as a series of arrested moments.

In a much-published photograph of Maison La Roche-Jeanneret (Figure 1.19) (designed with his cousin Pierre Jeanneret), the double-height gallery with mezzanine is shown in two-point perspective from a station point on the upper level. In it Le Corbusier was perhaps trying to represent pictorially the temporal qualities formulated in his discursive representation of the project. While the plethora of diagonals and overlapping planes in this iconic photograph beg an analogy with a Cubist painting, this composition is certainly not as sophisticated or assured of its intentions as those of the photographs discussed above. Unlike the others, this photograph was composed using two-point perspective. As we have seen, one-point perspective stresses frontality but permits variable readings of depth. Certain vantage points deliver ambiguous readings of space; others create tension between deep and shallow, convergent and frontal. The one-point perspective represents the numerous locations where the viewer places himself or herself in registration with the architecture, where a clear focal point in a frontal plane will visually organize the elements around the axis between the viewer and the focal point.

While it may provide more, and less ambiguous, information, a two-point perspective does not afford the registration between the subject and the architecture. If the station point is not determined by a well-defined visual axis, then the emphasis for devising the composition will be solely on the properties of the object. Rather, like the omniscient narrator in literature, the observer can be everywhere at once. But Le Corbusier's use of the second person in his description of this house restricts the viewer to a certain realm of probability and the reference to developing perspectives suggests that there are moments in which the views will be more resonant than others. Clearly, he is not unaware of the presence of the subject in his architecture and in this photograph as well there exist traces of the viewer. First, there is the fragment

Figure 1.19 Le Corbusier. Paris: Villa La Roche 1923 © 2003 Artists Rights Society (ARS)/ ADAAGP, Paris/FLC L2(12) 74.

of the mezzanine rail in the front left foreground, which establishes the location of the subjective viewer on the same mezzanine. Without that fragment the person or camera lens occupying the station point would appear to float in air. Second, one of the two vanishing points is still within the frame of the image. While not as resolute as the focal points in the later photographs, the window on the stair landing does "anchor" the subject's view. Third, Le Corbusier's image contains several possibilities for co-location, places where the viewer can conceive of being. Movement, and thus time, are introduced into a single image. If the subjective viewer is located in the image then so, too, are that viewer's "motor projects." Conceiving of movement through the space, for example, the itinerant viewer, on relinquishing the station point, must first negotiate the vertical void by moving out of the image to the left, then return along the mezzanine at left, disappear behind the opposite wall of the gallery and arrive at the projecting balcony at the top of the stairs. The subject interacts with the surrounding architectural object. From the new location, the subject may register as a new focal point the place where he or she has been; the image is reversed. This subject is not a ubiquitous observer but simply one who has assumed a new station point. From the new station point new routes will be detected and embarked on. The subject's station point, though fixed in the photograph, can be conceptually altered by the subject.

Antonio Martinelli's provocative photograph of Carlo Scarpa's Plaster-Cast Gallery in Possagno, near Carrera, imparts mystery to the idea of co-location (Figure 1.20). The photographer locates himself (and his camera) at a station point where he is in "registration" with a series of spaces in enfilade while he is simultaneously able to apperceive another space that denotes an alternative path through an adjacent gallery. The photograph is, in fact, in two-point perspective; it is only the convention of depicting spaces in enfilade in one-point perspective and the strong focal point at the end that could make us assume otherwise. As in Le Corbusier's representation of the gallery at Maison La Roche-Jeanneret, one of the vanishing points is clearly within the frame of the image, at the approximate end of the enfilade, where a sculpture is the focal point.

Figure 1.20 Carlo Scarpa, The Plaster Cast Gallery, Possagno. Photograph by Antonio Martinelli.

Everything I see is in principle within my reach, at least within reach of my sight, and is marked on the map of the "I can." Each of the two maps is complete. The visible world and the world of my motor projects are each total parts of the same Being....

Maurice Merleau-Ponty, "Eye and Mind"

But unlike the juxtaposed focal points in Le Corbusier's deep space/shallow space photographs, both of which are in conceptually accessible frontal planes (even if hidden from view), the focal point of the adjacent gallery does not appear to be at the end of a literal visual axis with the viewer, but rather at the end of a visual axis where the viewer *could be*. This mysterious focal point cannot be discerned without the viewer's movement. The viewer may conceive of himself or herself moving to and occupying either location; both are on Merleau-Ponty's map of the "I can." One is clear, the other enigmatic; but both are equally intriguing and equally command the viewer's movement. The irony of this brilliant photograph is that, while there are clearly two focal points for the viewer, one culminating each gallery sequence, the perspectival vanishing point for both is the same.

It is impossible to know whether Scarpa planned the occurrence of this visual event. A surface study of the body of his architectural work affirms that he deliberately employed the spatial enfilade, and the many instances of photographs based on obvious focal points make it difficult to refute that Scarpa consciously and compellingly exploited the straightforward visual effects of traditional architecture. But did he consciously consider the visual effect of this particular spatial juxtaposition? Or is it simply a case of the aleatory – the discovery by the viewer/photographer of something that occurs by chance as a result of other intentions and operations?

The spatial experience of parallax, or perspective warp, while moving through overlapping spaces defined by solids and cavities opens the phenomena of spatial fields. The experience of space from a point of view that is in perspective presents a coupling of the external space of the horizon and the optic point from the body.

Steven Holl, Anchoring

Subjective perception conjoined with subjective movement multiplies the visual effects that are captured, frozen, in the typical representation. Any representations of this experience must include a multitude of views, with a multitude of focal points, each registering with the surroundings in a different way and each disclosing different readings of depth. An objective representation in time would scan the object without registration, without determinant focal points and without any consequential readings of depth. Subjective movement introduces another effect which can only occur in its presence – parallax. Originally a term used in astronomy, the dictionary definition of "parallax" is "the apparent displacement of an observed object due to a change in the position of the observer."[16] An objective representation of parallax is impossible.

More than 30 years after Le Corbusier wrote about the architectural promenade, he designed the Dominican monastery at La Tourette. Colin Rowe's narration about the approach to La Tourette (1961), a fragment of which appears at the beginning of this chapter, is premised on his movement as he approaches the monastery. As he changes his position, focal points displace one another, figure and field fluctuate; his forward movement is first impelled and then repelled. The dominant motif of Rowe's perceptual experience is the change in the relationships between things he sees as he moves through the visual field. The meanings with which he imbues these changes are, of course, personal, and it is only with unusual temerity (for Rowe) that he broaches the possibility of intersubjectivity when he proposes that Le Corbusier might have contrived this experience for the viewer.

This extraordinary overlapping, which we never think about sufficiently, forbids us to conceive of vision as an operation of thought that would set up before the mind a picture or representation of the worlds, a world of immanence and of ideality. Immersed in the visible by his body, itself visible, the see-er does not appropriate what he sees; he merely approaches it by looking, he opens himself to the world.

Maurice Merleau-Ponty, "Eye and Mind"

Rowe and Slutzky's essay "Transparency: Literal and Phenomenal," previously discussed with reference to depth which is visible in two dimensions, turns in its final pages to another work of Le Corbusier and Pierre Jeanneret, the competition entry for the League of Nations in Geneva – a three-dimensional version of phenomenal transparency. In this case Rowe and Slutzky had to hypothesize the subjective experience of the complex site plan because this work was never built (Figure 1.21).

The League of Nations programme required a large volume for the Secretariat, but the architects mitigated its size and singularity by producing a site plan in which the "highly assertive" deep space of the vast entry court is:

repeatedly scored through and broken down into a series of lateral references – by trees, by circulations, by the momentum of the buildings themselves – so that finally, by a series of positive and negative implications, the whole area becomes a sort of monumental debate, an argument between a real and deep space and an ideal and shallow one.

(Rowe and Slutzky 1997, p. 174)

There is no single overview of the great entry court. Instead, the viewer becomes aware of these lateral scorings, or striations, by moving along an axis

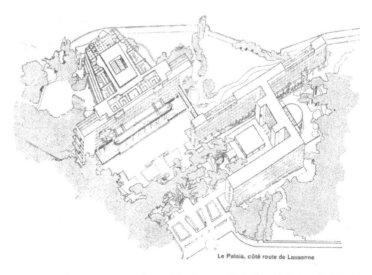

Le Palais, côté route de Lausanne

Figure 1.21 Le Corbusier. Genève: Palace of the League of Nations 1927 Plan FLC 23185 © 2003 Artists Rights Society (ARS)/ADAAGP, Paris/FLC.

that cuts through and past planes – of trees, of a terrace, of the narrow building blocks. The changing relationships of these elements can only be apperceived when the viewer is in motion. A screen of trees, for example, is discerned as a plane when it first "intersects" the viewer's vision, but then becomes simultaneously the lateral edge of one space and parallel edge of another when the viewer has moved beyond it. Again, Merleau-Ponty's comments on depth resonate: "[B]y virtue of depth [things] coexist in degrees, they slip into one another and integrate themselves" (1968, p. 219). The viewer's movement sets all this in motion. As he or she moves, depth and scale fluctuate, objects displace each other and reappear. So, too, do the emergent meanings of these varying relationships between the things and the subject. Parallax requires a subject and it requires time; it is "an intertwining of vision and motion."[17] What are the possible means of representing both the embodied viewer *and* time?[18]

The hyper-real experience of the League of Nations had to be surmised by Rowe and Slutzky largely on the basis of the axonometric drawings. While perspective drawings were included in the competition documentation, they were composed only to describe the objective properties of the project, with the station point being a physical impossibility unless the viewer was suspended in mid-air by a crane. In short, there exist no representations of the League of Nations project that are intended to convey the experience that Rowe and Slutzky describe.

Le Corbusier's perspective sketches often delineate spaces with discontinuous wall surfaces and freestanding elements such as columns (verticals), tables (horizontals), and stairs and ramps (diagonals) which, at least conceptually, allude to the experience of parallax which will be induced by the viewer's movement (Figure 1.22). But it requires an effort of imagination, such as Rowe and Slutzky's, to extract temporal experience from such an image. Steven Holl has experimented with paired representations that more literally speak of the changing relationships between architectural elements when the subject's point of view changes (Figure 1.23).

Depth . . . is the dimension in which the thing is presented not as spread out before us but as in inexhaustible reality full of reserves.
Maurice Merleau-Ponty, Sense and Non-sense

Holl's watercolours are more enticing, more suggestive of a multiplicity of views and changing relationships between things. The presence of the viewer is always implicated by the eye-level station point and the presence of a strong focal point, even when it is hidden from view (Figure 24). From the single point of view that suggests a primary path, there exist multiple spaces of

Figure 1.22 Le Corbusier. Neuilly sur Seine: Villa Meyer 1925 Plan FLC 31514 © 2003 Artists Rights Society (ARS)/ADAAGP, Paris/FLC.

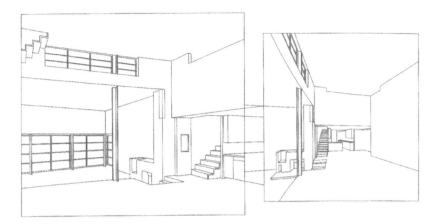

Figure 1.23 Steven Holl, Cleveland House, 1988. Courtesy Steven Holl Architects.

Figure 1.24 Steven Holl, Museum of Contemporary Art, Helsinki, 1998. Courtesy Steven Holl Architects.

co-location. Peripheral interest abounds as patches of light and dark allude to partially hidden spaces that complete themselves beyond the frame, not only to the left and right but also above and below (Figure 1.25). These perspective views do not represent a detached piece of the objective world, but rather a momentary framing or registration of the subject's vision within the continuity of the objective world. Further, the overlapping and folding of planes, the exploded surface of volumes, and the presence of ambiguous objects postulate the "inexhaustible reality" of parallax – the changing relationships between things as the viewer moves through the visual field (Figure 1.26). A discursive description inspired by these images may prove to

Figure 1.25 Steven Holl, Cranbrook Institute of Science, Bloomfield Hills, 1998. Courtesy Steven Holl Architects.

Figure 1.26 Steven Holl, Stretto House, Dallas, 1991. Courtesy Steven Holl Architects.

be an enjoyable invention, but the "wordlessness" of these images will always be more provocative.

Bernhard Berenson[19] in his book *Aesthetics and History* admonished art historians and theorists against failing to differentiate between the producer and the consumer of a work of art. This search for the presence of the subjective viewer in design representation has concentrated on analyzing the artefact rather than its role in the design process. It has been difficult, however, to avoid all allusion to the implications of this analysis for the genesis of design ideas. In the absence of specific statements of intention by architects, it is necessarily speculative to assume there are any implications at all. Could it be, as Colin Rowe conjectures about Le Corbusier, that an architect would characterize his or her design process as being "obedient to the exigencies of the eye rather than those of the work, to the needs of the conceiving subject rather than the perceived object"? Could it be, as Maurice Merleau-Ponty suggests, that "The joy of art lies in its showing how something takes on meaning – not by referring to already established and acquired ideas but by the temporal or spatial arrangement of elements" (1964, pp. 57–58)?

The questions will remain open, but the speculation itself raises another question. What of our tools of representation? To the extent that such analysis *could* inform the design process (and the current milieu of cultural criticism suggests that it might), a critique of the tools of representation that are available to abet such design thinking should not be avoided. It is no longer legitimate to deny the ubiquity of computer applications as the primary tool of representation for the architectural designer. Although there are a multiplicity of applications, the focus of their performance is on the description of the object and the means of its transformation. While distances can easily be manipulated, these applications are not readily suited to altering objects with respect to subjective views – for example, of depth. No more are they prepared to represent the human experience of objects in a delimited space with anything like Merleau-Ponty's world of the "I can."

We get from our tools of representation only what we ask of them. To reaffirm the value of human experience, to reinstate the human subject in the world, we may wish to ask for or invent tools that include the presence of the subject in design representation and therefore in the design process.

Notes

1. For example, see Harbison, R (1997) *Thirteen Ways*, The MIT Press, Cambridge, MA, 160–173.
2. The use of the word symbol here is a direct reference to Ernst Cassirer's concept of "symbolic form," which derives from Kant's notion of categories.
3. Michael Ann Holly's research on Erwin Panofsky's early essays demonstrates the extent to which Panofsky attempted to understand art history synchronically but was nevertheless unable to avoid the diachronic aspects of his own contemporary philosophical influences. Holly, M.A. (1984) *Panofsky and the Foundations of Art History*, Cornell University Press, Ithaca.
4. The metaphysical and epistemological basis for the modern view of subject and object must be understood in post-Kantian terms. Whereas the existence of the objective world is reaffirmed, *knowledge* of it can only be subjective. While Kant does not eradicate the mind–body dualism, he does not assume, as Berkeley did, that all matter exists only in the mind, or, as Descartes did, that the properties of matter are subjectively constituted.
5. Merleau-Ponty makes such a distinction when discussing Bishop Berkeley's argument for the necessity of intellectual spatial restructuring of the visible world. Merleau-Ponty, M. (1996) *The Phenomenology of Perception*, p. 255.
6. My review of work that could be considered generative was necessarily cursory and mostly in vain. It is, I believe, well worth extended research. Such research is made difficult, however, because of what Lauretta Vinciarelli (Colomina, p. 245) has identified (in contradistinction to Croset's loss of the experience of the building) as the "loss of the experience of the project." With the exception of the publication of sketchbooks of a few major architects, there is little published documentation of drawings that reveal the design process.
7. Computer-generated animations of architectural designs are notorious for this disembodiment: gravity-defiant "cameras" often produce fly-throughs or crawl-throughs instead of walk-throughs.
8. Edgerton, S. (1975) *The Renaissance Rediscovery of Linear Perspective*, Basic Books, New York, p. 26.

9. Evans offers collaborative photographic evidence to suggest that Mies was well aware of this effect.
10. According to Samuel Edgerton, Alberti's method for determining what is now known as the distant point was the most consistently ignored of Alberti's rules for constructing the image. Edgerton, S. p. 55.
11. Alberti's "centre point" is now generally referred to as the "vanishing point" because of its theoretical association with a point at infinity. It is interesting to note that for more than two centuries pictorial perspective was practised without such knowledge or association. It was only in 1639 that Girard Desargues observed the mathematical connection. Pérez-Gómez, A. and Pelletier, L. (1997, pp. 132ff).
12. See, for example, Jay, M. and Crary, J. (1992) *The Techniques of the Observer: On Vision and Modernity in the Nineteenth Century*, The MIT Press, Cambridge, MA; Levin, David Michael (1993) *Modernity and the Hegemony of Vision*, University of California Press, Berkeley.
13. No doubt influenced by the notation of a geometric centreline that was essential to Beaux-Arts parti studies, he severed himself from the academic tradition of planning which produced nothing but planimetric "star shapes." Le Corbusier (1968, pp. 180 and 191).
14. For purposes of comparative clarity, I am continuing the use of the term "vanishing point" even though it is an anachronism with respect to Piero's painting.
15. In fact, initial analysis would suggest that Piero's choice of the vanishing point is not a primary compositional device, even though he uses it instrumentally to establish the "deep space." Robin Evans' analysis of this painting consults Piero's own *Treatise on perspective* to suggest that he was employing more than one method of projective construction in this painting. Evans, R. (1995) *The Projective Cast: Architecture and Its Three Geometries*, The MIT Press, Cambridge, MA, 147ff.
16. (1979) *The Random House Dictionary of the English Language*, edited by Jess Stein, Random House, Inc., New York.
17. Merleau-Ponty, M. (1954, p. 162).
18. Perhaps an obvious answer would be the cinematic art forms, but they, too, are primarily in the service of rendering images objectively rather than subjectively. Animation, in particular, must be objectively conceived and executed before it is set in motion.
19. Berenson, B. (1954) *Aesthetics and History*, Doubleday, Garden City, NY.

References

Alberti, LB 1991. On painting. London: Penguin Books.
Baudrillard, J 1988. Simulacra and simulations. In Jean Baudrillard: Selected writings, edited by M. Poster. Stanford: Stanford University Press.
Croset, P-A 1988. The narration of architecture. In Architecture production, edited by B. Colomina. New York: Princeton Architectural Press.
Evans, R 1997. Translations from drawings to buildings. Cambridge, MA: MIT Press.
Giedion, S 1967. Space, time and architecture: The growth of a new tradition. Cambridge, MA: Harvard University Press.
Hoesli, B 1997. Commentary. In Transparency, edited by C Rowe and R Slutsky. Boston: Birkhauser.
Holl, S 1980. Alphabetical city: Pamphlet architecture No. 5, San Francisco.
—— 1991. Anchoring. New York: Princeton Architectural Press.
Jameson, F 1983. Postmodernism and consumer society. In The anti-aesthetic: Essays on postmodern culture, edited by H Foster. Seattle: Bay Press.
Jay, M 1994. Downcast eyes: The denigration of vision in twentieth-century French thought. Berkeley: University of California Press.
Kepes, G 1944. Language of vision. In: The Mathematics of the ideal villa and other essays, edited by C. Rowe. Cambridge, MA: MIT Press, p. 11.
Le Corbusier and Jeanneret, P 1935. Oeuvre complète, vol. 1. Zurich: Editions Girsberger, p. 60.
—— 1968. The modulor: A harmonious measure to the human scale universally applicable to architecture and mechanics. Cambridge, MA: MIT Press.
—— 1986. Towards a new architecture. New York: Dover Publications. Reprint.
Levin, DM 1988. The opening of vision: Nihilism and the postmodern situation. New York: Routledge.
Merleau-Ponty M 1954. Eye and mind, translated by Carleton Dallery. In: The primacy of perception and other essays on phenomenological psychology, the philosophy of art, history and politics. Evanston: Northwestern University Press.

—— 1964. M. Sense and non-sense. Evanston: Northwestern University Press.

—— 1968. The visible and the invisible. Evanston: Northwestern University Press.

—— 1996. Phenomenology of perception. London: Routledge.

Panofsky, E 1991. Perspective as symbolic form. New York: Zone Books.

Pérez-Gómez, A and L Pelletier 1997. Architectural representation and the perspective hinge. Cambridge, MA: MIT Press.

Rowe, C 1976. The mathematics of the ideal villa and other essays. Cambridge, MA: MIT Press.

Rowe, C and Slutzky, R 1976. Transparency: Literal and Phenomenal. In: The mathematics of the ideal villa and other essays, edited by C Rowe. Cambridge, MA: MIT Press.

Shumacher, T 1987. Deep space, shallow space. Architectural Review January 1987.

Graphic Representation as Reconstructive Memory: Stirling's German Museum Projects

Gabriela Goldschmidt and Ekaterina Klevitsky

Introduction

Between 1975 and 1977, James Stirling's office[1] entered three competitions for the design of museums in Germany. Although only the last design, for the Staatsgalerie in Stuttgart, was a winning entry that resulted in the building of the museum according to Stirling's plans, all three designs were widely published in leading architectural magazines. Of particular interest are the publications of the designs for the Nordrheine-Westfalen Museum in Düsseldorf and the Wallraf-Richartz Museum in Cologne in *The Architectural Review* in 1976[2] and even more so in *Lotus International* in 1977.[3] What made these publications unusual was the inclusion of types of representation that were not commonly encountered in contemporary architectural publications: a large number of axonometric line-only drawings and small freehand sketches. The article in *Lotus* ended with a graphic collage that put the Düsseldorf and Cologne design principles into a broader context of Stirling's work. Various projects and the links among them were demonstrated by the same kind of analytical drawings, as the architects called them. The Staatsgalerie in Stuttgart, the third in Stirling's "German museum trilogy," was presented in *Architectural Design*[4] in late 1977 through photographs of the building's model and an array of typical small-size Stirling "doodles." Axonometric drawings of the Stuttgart museum were published elsewhere later on.[5] Figure 2.1 displays a sample of sketches made for the Düsseldorf museum. A sheet of sketches made for Stuttgart is reproduced in Figure 2.2.

The "axos," as the axonometric drawings were called, were partial and "edited" views of selected elements of

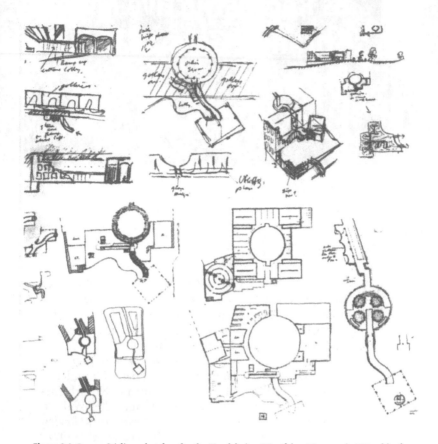

Figure 2.1 James Stirling, sketches for the Nordrheine-Westfalen Museum in Düsseldorf.

the proposed buildings, rather schematic, as opposed to habitual detailed and "rendered" views of buildings that were the norm in publications (i.e., using texture, shading etc. to "enliven" the representation). The axonometric drawings were quite abstract and could hardly be taken to represent views of the building as it would be perceived by an observer, regardless of his or her station point. Rather, the architects were interested in exposing the major concepts that guided the design in as "pure" a manner as possible. The reductionist strategy they adopted and the choice of impossible viewing angles served the purpose of explaining the concept instead of presenting the building. Along with other images that supported the representation of design concepts, we shall refer to the drawings in the competition publications as "conceptual drawings".

The published drawings in the various reports on the German museums were a unique mixture of conventional presentation drawings (plans, sections, and elevations), conceptual drawings (mostly schematic axonometric views) and preliminary freehand sketches. Some were exploratory drawings made during the process of design; others were "after" drawings, made post factum to convey particular images and concepts: the ideas and principles that went

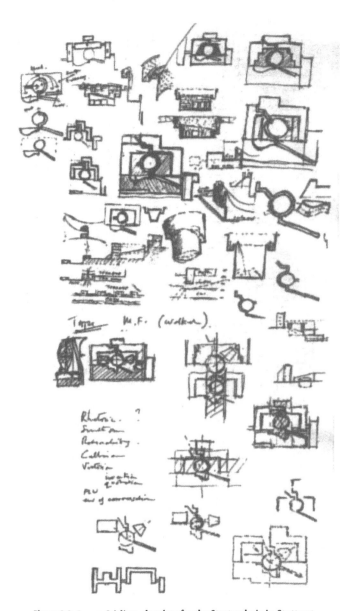

Figure 2.2 James Stirling, sketches for the Staatsgalerie in Stuttgart.

into the making of the building. Figures 2.3, 2.4, and 2.5 – which pertain to the Wallraf-Richartz Museum in Cologne and present a standard ground plan drawing, an abstracted top-down axonometric hard-line drawing and a sketch axonometric drawing of the scheme respectively – serve as examples of the various types of drawings. Models never played an important role in Stirling's design processes (Stirling 1992) and, therefore, for the most part, their photographs are not prominent in representations of the projects.

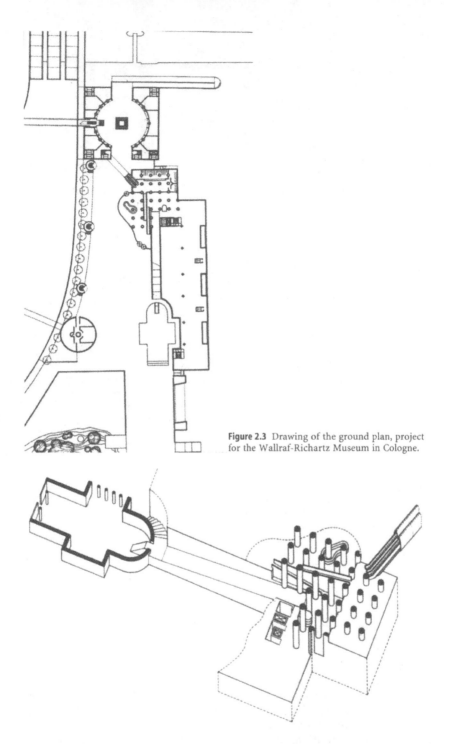

Figure 2.3 Drawing of the ground plan, project for the Wallraf-Richartz Museum in Cologne.

Figure 2.4 Top-down axonometric view of the Wallraf-Richartz Museum in Cologne.

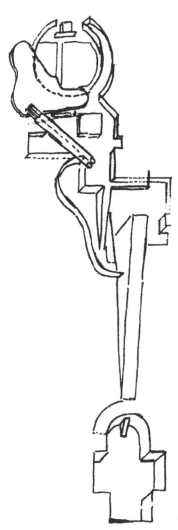

Figure 2.5 Sketch, axonometric drawing for the Wallraf-Richartz Museum in Cologne.

In this chapter we look at the particular mixtures of drawings that were selected for printing when Stirling's museum designs were first made public. We treat the series of drawings as pictorial design narratives, each comprised of a carefully composed collage of images. We maintain that these representations had a forceful impact and served as forerunners of innovative modes of architectural representations in later years. At the same time we ask what contemporary cultural trends nurtured the kinds of images that Stirling opted to avail himself of and what made them acceptable to the public. We conclude that a newly found interest in the design process and in the ideas and concepts that guide it, led to the wish to make public not just the resulting artefact, but also the narrative that tells the story of the process and concepts as well. We use the term "reconstructive memory" to describe the unconventional representation that mixed together standard drawings of the buildings with evidence from the preliminary conceptual search, together with abstractions made post factum, in order to tell the design story.

The Context

James Stirling – a profile

James Stirling was born in 1924. After the Second World War, during which he was enlisted, he studied at the Liverpool University School of Architecture where one of the very young professors was Colin Rowe. (We shall return to Rowe's influence on Stirling further on.) He started practising in the early 1950s and, until his death in 1992, had been in private practice by himself and with partners (from 1971 with Michael Wilford), while also teaching in England, Germany, and the USA. Stirling was born into Modernism: Wilson (1992) points out that two of Modernism's "archetypal masterpieces" (Le Corbusier's Pavilion de l'Esprit Nouveau, and Byvoet and Duiker's Zonnestraal Sanatorium in Hilversum) were built shortly after Stirling's birth. As a young architect Stirling was, like most of his contemporaries, an avid Modernist who saw himself as a disciple of Le Corbusier. His first significant large project was the Leicester University Engineering Building, designed in 1959. By the time the Cambridge University History Faculty Building was completed in 1967, Stirling was already a very well-known architect. In the mid-1960s it became obvious that Stirling was no longer an orthodox Modernist. He continued to base his designs on rational analyses of programme and context, but his forms became less constrained and "boxy." Instead of subdividing space within a prismatic volume as in Le Corbusier's "plan libre," he started assembling independent spaces, enclosed in distinct volumes, around flexible circulation spaces – both horizontal and vertical. This was a combinatorial act that yielded elaborate forms that were joined together with great mastery. While still manifesting an interest in the work of some of the pillars of Modernism which included, in addition to Le Corbusier, Kahn, Aalto, and others, Stirling did not refrain from studying pre-Moderns like Asplund and 19th-century neoclassicists like Schinkel, for example. Stirling, who cultivated eclecticism, never shied away from "borrowing" forms he liked: "Like Picasso, Stirling operated a magpie avidity to steal whatever he liked while yet turning it into his own – and that is a freedom which is only possible to someone who belongs to no school" (Wilson 1992, p. 20).

Despite Stirling's strong individuality and the various freedoms he took in designing, nonetheless he did work in a very consistent way, both in terms of the ideas he pursued and the design searches he conducted. He testified about his design principles: "I never think of a design as being conceived from the outside; on the contrary, all our designs are conceived following the sequence of entry and going through primary movement" (Stirling 1992, p. 24). By "primary movement" he meant the main circulation spaces of the building or complex of buildings. Organizing buildings on the basis of circulation was, of course, a fairly common concept. Stirling took it to an extreme and, in particular, he allowed spaces devoted to circulation, like corridors and staircases, to occupy primary forms in his designs, equal in importance and elaboration to spaces devoted to "useful functions." The use of circulation, or movement, as the generator of form is evident in Stirling's work from the beginning of his career (Jacobus 1975). When his buildings began to lose their compact boxiness in favour of assemblages of individually crafted volumes, this principle became forcefully apparent. An example is the residential expansion for St Andrews University (UK), completed in 1968, where the scheme is one of long "fingers" of corridors that give access to students' rooms in a herringbone pattern.

In the process of preliminary design, developing forms was a clear top priority with Stirling. Form was never arbitrary; rather, it was developed piecemeal "in response to the site and to major factors such as entry and procession through the building." Decisions about structure, materials, colours and more were made "when the process is well under way" and "the entire concept is worked out" (Stirling 1992, p. 24). Despite the primacy of three-dimensional form, the design search was conducted in two-dimensional media only – namely, drawings and not models (which were added when "the design is over," for presentation to journalists, planning authorities or clients). Drawings, however, were always made in abundance.

In the first phase of design Stirling used to make a large number of very small sketches: "the drawings which accompany this sort of thinking are doodles, tiny sketches about one centimeter in size" (ibid., p. 24). Stirling made them on every available piece of paper; some were quite literally "back of envelope" and "cocktail napkin" sketches made in transit; other sketches filled many a sheet of paper and were executed in the office. Co-workers remember Stirling as "always doodling, doodling" (Girouard 1998, p. 188); he "constantly searched for ideas in diagrammatic sketches" (Livesey 1992). The small "diagrammatic sketches" could consequently "be developed in small axonometric drawings showing the relationships between volumes and heights" (Stirling 1992, p. 24). At that point Stirling was ready to involve his assistants in the process: "my doodles give the lead, then a sort of tennis match begins with my colleagues" (ibid., p. 24). This "tennis match" was described by one of those colleagues as follows: "The office was a searching factory for an investigative architecture. It was not a pick and choose operation, but a constant elaboration of an agreed on set of ideas, a sequence of sketches rather than 25 alternatives. . . . The process remained absolutely consistent from beginning to end: collaged incidents on a set of overlaid formal arrangements" (Livesey 1992, p. 70). Many of these drawings were axonometric views: "The axonometric was promoted by Jim as part of his design process and, as has been pointed out by others, the design of a building like Leicester would be unimaginable without it. . . . These were all of the "Modernist" down-view variety. It was not until Leon Krier joined the office later in 1968 that the edited Choisy up-view joined the repertoire" (Jones 1992, p. 70).

Jones' remark concerning axonometric down-views and up-views requires explication. Axonometric views, for which the technique had been known since the Renaissance, penetrated architectural representational conventions in the 1920s as partial fulfilment of Modernism's need for new tools to express new ideas and concepts (Bois 1981). Commonly, modern architects drew down-views or "bird's-eye" views – representations that look down at the depicted object. As noted above, Stirling made very frequent use of such axonometric views, both in his sketches and in the hard-line drawings the office produced (see, for example, Figures 2.4 and 2.5). Krier was hired to make drawings for Stirling's book (1975); he redrew many of the older projects in order to create a clear and unique image of Stirling's work. While at it, he also initiated the second brand of axonometric views, up-views or "worm's-eye" views, unconventional representations that look up at the depicted object (Girouard 1998). As Jones (1992) pointed out, the up-views were revivals of Choisy's schematic axonometrics that offered, towards the end of the 19th century, a rationalist construction-centred view of architecture. The up-views were an uncommon, if not unique, mode of representation in the 1970s because they depicted a virtual view from an impossible angle and therefore did not really give information about the looks of the depicted building. But this was precisely the reason they were so appealing to Stirling: they enabled the representation of a concept, a scheme, in a diagrammatic manner. Unlike Choisy, an architect/engineer who used up-views to represent the structural and constructional principles of a building, Stirling used them for other purposes: he was interested in showing the essentials (and only the essentials) of the relationship between form, space and movement. This choice of means turned out to be so appropriate in terms of Stirling's intentions and the image he wished to project that up-views became a standard stock of his office, almost its trademark. Figures 2.6 and 2.7 show up-view axonometric drawings prepared for the Düsseldorf and Stuttgart competitions, respectively.

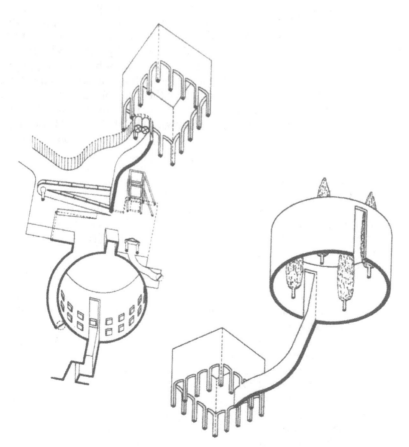

Figure 2.6 Axonometric up-views of major elements in the Düsseldorf Museum design.

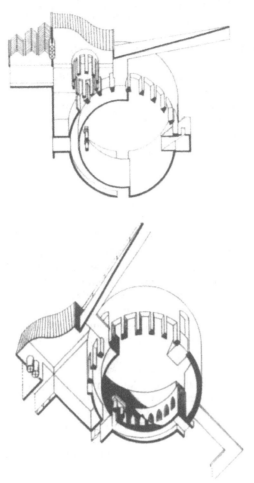

Figure 2.7 Axonometric up-views of major elements in the Stuttgart Museum design.

Postmodern representation

The 1970s are often referred to as the decade of the emergence of postmodernism. The architectural critic Charles Jencks (1982) lists the Stuttgart museum by Stirling as one of the major examples of postmodern Classicism, which, he claims, is "one half of the style toward which Post Moderns turn" (ibid., p. 12). The other half grows out of Late-Modernism and its style is marked by a desire for richness of form, achieved through compositional play and dissonance of elements. Skewed grids, violent juxtaposition of volumes and exaggerated solids, voids and building components are some of the means used to attain the wealth that architects who subscribed to this style aspired to. Postmodernism is, however, pluralistic and eclectic; purity of style is not one of its values and no sharp borderlines separate between these twin postmodern styles. Tzonis and Lefaivre (1992) see the "spring of 1968" as the decisive point in time that marks the beginning of a new era in Europe, inclusive of its architecture. Among the important characteristics of the architecture of the two decades following the 1968 events, they name populism, neo-rigourism, skin rigourism, and the call to disorder. By neo-rigourism they mean the "true" expression of functional and structural aspects of buildings. Skin rigourism alludes to a certain "emancipation" of the surfaces of which the envelope of the building is made in terms of composition, materials and detailing. The treatment of the envelope, in this view, is no longer in the exclusive service of the functions embodied in the volumes it encloses. Instead, it can be made of independent elements that have their own intrinsic goals – aesthetic, technological and otherwise. The call to disorder, where it is manifest, is a need to break with order as representative of old times, a desire to escape from the past and facilitate the emergence of a new world view. Tzonis and Lefaivre (ibid.) point to past "periodic eruptions of love for disorder and a desire to escape from the ideas of coherence and system of the type . . ." (ibid., p. 20). An example they give is constructivism, with its architectural and artistic manifestations in the 1920s in Russia.

In *The History of Postmodern Architecture*, Klotz (1988) proposes that "the characteristic objective of postmodernism – [is] to create an architecture of 'narrative contents'" (ibid., p. 128). He sees the pursuit of this objective as a reaction to the "radical functionalism" dogmas of Modernism. In Klotz's words, postmodernism "deflects one's attention [from "the bare factuality of architecture"] to the completely different realms of environment as a narrative representation" (ibid., p. 128). The notion of narrative representation is one we will return to shortly. Klotz sees the postmodern programme as an attempt to liberate architecture from the "radical abstraction" that marked "the architecture of functionalism" and which was "something dry, rigid and lacking freedom." The latter pejorative characterization of abstract architecture is taken from Karl Friedrich Schinkel who was dissatisfied with the architecture of his day (early 19th century) for similar reasons to those that motivated the "rebellion" of postmodern architects in the 1970s and 1980s. As already pointed out, Schinkel was, among others, a source of inspiration for James Stirling, especially in his work on the Düsseldorf and Stuttgart museums.

The writings of Jencks, Tzonis and Lefaivre, and Klotz, are faithful representatives of literature about postmodern architecture. They expose and analyze roots, trends and intentions in one of the most significant periods of transition in the architecture of the modern era. However, they do not address the question of the representation of this architecture: the representational means with which architects carried out their design explorations and presented them to themselves, their colleagues and to the public at large. It is true that transformations in graphic means of representation were nowhere as dramatic as the changes we witness in the architecture they convey. Conventions of graphic representation, as established during the Renaissance, have served designers well for centuries and continue to do so as we speak. But if we look a little closer we will find that, within and around the basic modes of representation that are embodied in orthogonal projections and perspectives, a wealth of variations, combinations and novel representational types have appeared over time.

Innovations in representational modes are neither random nor arbitrary phenomena. We propose that they come into being when powerful new design concepts seek a path to the consciousness of viewers. Klevitzky (1997) has shown how the abstract three-dimensional compositions that were practised in the Bauhaus and Vkhutemas schools in the 1920s were a response to a novel philosophy of architectural and design education. This philosophy emerged as part of the avant-garde movements in the visual and performing arts and all design disciplines, as well as from new ideas in psychology and education. Postmodernism, with its new tenets, required a lesser revolution: according to Klotz (1982) its emergence enjoyed a "smooth transition" from modernism. All the same, there were architects who were dissatisfied with "conventional" means of representation and found it necessary to display their work somewhat differently in order to express its meaning fully. The publications of Stirling's German museum projects fall into this category.

It is interesting to follow architectural publications and detect subtle changes over time in the nature of drawings presented in them. We can afford only a cursory sample in this chapter, but even this brief glance is revealing. In Gössel and Leuthäuser's *Architecture in the Twentieth Century* (1991), freehand sketches by Mendelsohn and Le Corbusier, both from the late 1920s, are the earliest exemplars of this type of representation. Likewise, the first axonometric drawing is of a late 1920s project by Gropius. Unlike the sketches, which are "on-line" documents made during the process of designing, the Gropius axonometric and others that follow are "after the fact" drawings, made for explanatory purposes. We see the rise of interest in sketches and axonometrics in the 1920s as part and parcel of the new architectural programme of the modern movement (Bois 1981; Klevitzky 1997). The next axonometric drawings we encounter in Gössel and Leuthäuser (1991) are from the late 1960s (work by Stirling and by Hejduk) and the 1970s and 1980s, two decades that contributed a fairly large number of axonometric drawings and some freehand sketches to this overview of architectural work. We maintain that it is not by accident that we encounter sketches and axonometric drawings in the 1920s and then again in the 1970s, but rarely in between. The new postmodern aspirations of architects in the 1970s gave rise to a need for new representational means similar to, if less forceful than, the needs of avant-garde architects of the 1920s. When anthologies of postmodern architecture began to be published around 1990, the trend was already clear and we therefore detect it in periodicals and books. An interesting and unusual example is a book titled *100 Contemporary Architects: Drawings and Sketches* (Lacy 1991) in which all but a few drawings are in fact rapid freehand sketches. (Stirling is represented in this book by two sheets of doodles made while designing the Clore Gallery at the Tate Museum, London, in 1986.) Never before has there been so much interest in "doodles" made during the process of architectural design.

In the mid-1970s, however, this trend was still all but non-existent. Sketches were made, of course, but were considered a private matter, a rough exploratory tool, of interest only to the exploring designer and of no value once the design has progressed to an advanced stage in which hard-line drawings are made. Even Stirling treated his sketches this way; many of his early sketches were eventually discarded as having no value at all (Girouard 1998; Wilford 2000).[6] Therefore, the 1976 and 1977 publications of sketches and axonometric drawings by Stirling may be considered a pioneering act, a swallow that heralded the spring.[7] This is the reason for our particular interest in these series of drawings and in their publication.

Principal Design Concept: Promenade Architecturale

Given Stirling's interest in movement through buildings ("all our designs are conceived following the sequence of entry and going through primary movements"), it is easy to understand his excitement about the design of museums, a building type that is traditionally regarded as evolving around circulation routes. Furthermore, the programmes of all three German museums called for the inclusion in the designs of public footpaths through the sites, connecting adjacent parts of the respective cities to one another. Participating in the museum competitions was therefore an opportunity to explore, in addition to the circulation through the museum buildings themselves, circulation patterns at an urban scale.

The programme for the Düsseldorf museum – the first of the three competitions – required that the design create a pedestrian path through the site along the routing of the old town wall, connecting two public squares on either side of the museum site. These squares, Grabbeplatz and Ratinger Mauer, had been linked in the past and this linkage was to be revived. At the same time, at a larger scale, the footpath through the museum premises was to facilitate connection to the old town and the integration of important existing buildings into a unified network of pedestrian paths. The Cologne competition was, in fact, more than a building design competition: the program explicitly requested an urban solution for a difficult site with a wide stretch of railway functions adjacent to its long side and with the Rhine and the famous Cologne Cathedral abutting the narrow ends of the site on either end. The railroad installation had cut off the northern town centre from the rest of city. The design of the museum and its environs was to help overcome this problem, among others, by a careful layout of pedestrian routes and enclosed and open public spaces. The Stuttgart site, where a major extension of the 1838 Staatsgalerie was to be designed along with a new Chamber Theatre, is in a district of cultural institutions on a hillside. The hill is separated from the city centre by a busy eight-lane highway at its feet, which was created as part of post-war urban renewal plans. The site included a number of existing buildings along local streets, the fate of which was to be determined by the designers. The expected merits of design proposals for the site clearly included an improved urban structure in that area, with a special emphasis on pedestrian connections.

Nothing could have suited Stirling better. In every one of the projects he created numerous public spaces, of which one was conceived as a key design element at the heart of a pedestrian circulation network within the site and from there leading into the city. In Cologne this was a combination of an entrance peristyle hall and a sculpture garden (see Figure 2.3), and in Düsseldorf and Stuttgart these were inner courts in the shape of circular drums: a garden in Düsseldorf and a rotunda in Stuttgart. Several critics (e.g., Curtis 1984) suggest that these cylindrical spaces were most likely inspired by Schinkel's 1825 Altes Museum in Berlin that surrounds such a space.[8] Stirling himself singled it out as a precedent: "I'd like the visitor to feel it 'looks like a museum ... I would refer to Schinkel's Altes Museum as representative of the 19th-century Museum as a prototype" (Mendini 1984). In Düsseldorf the round garden was complemented by, and directly linked to, a large pavilion, or kiosk, that marked the entrance to the museum from the Grabbeplatz. The ground plan of the complex is shown in Figure 2.8 (see also Figure 2.6). In

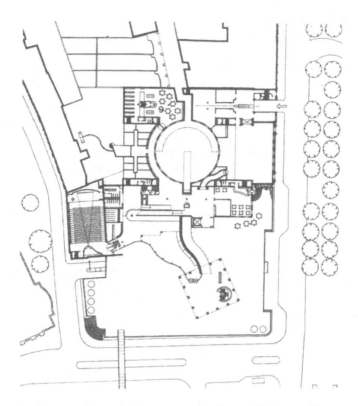

Figure 2.8 Drawing of the ground plan, project for the Nordrheine-Westfalen Museum in Düsseldorf.

Stuttgart the topography required a climb up the hillside and this climb was negotiated by a spiralling ramp along the walls of the circular courtyard or rotunda (see Figures 2.7 and 2.9). Although public, these spaces were very much part of the architectural design solution for the museum buildings themselves, thereby fusing public urban pedestrian movement with the museum-going experience. This theme was so powerful that in fact it became the main design concept of the three museum projects.

In accordance with this concept, Stirling was very concerned with the architectural elements that served as vessels for pedestrian movement: the path, the ramp, and their relationship to the open spaces they interconnected. Questions of open versus closed spaces, narrow and linear versus spacious and equi-dimensional places, and meandering as opposed to straight paths were investigated. The angle at which a path penetrated a square was important. Everything was geared to enrich the experience – primarily the visual experience – of those who proceed along the walkways or stop in the public spaces. At the same time the architects attached prime value to the formal compositional qualities that resulted from the combination of the various elements that lay alongside the circulation routes. In Corbusian terms, Stirling worked on the promenade architecturale that for him crystallized the major concept of each of these designs (Le Corbusier 1935; Baker 1992a, b).

Pictorial Design Narrative

When the time came to publish the designs, Stirling apparently felt that presenting final plans was inadequate. This was not a conventional building that could be represented through plans, sections and elevations. In fact, elevations were a tricky business – the Staatsgalerie, for instance, does not have a street elevation (facing the highway), so to speak. Stirling thought that for this new and dynamic urban complex that was all about movement, a different mode of representation was called for if he was to let the reader share, at least to some degree, the experience of the movement systems with all of their intricacy and the story of having created them. There was a story to tell, a narrative that was to be presented using pictorial images rather than words. As pointed out above, narrative was in good currency at the time: Klotz (1988) sees narrative representation as one of postmodernism's principal tenets. Stirling, however, appears to be the first architect who literally translated the narrative imperative into a series of images, with no particular sequence to them, that as an ensemble "collaged together" the desired narrative. In doing so he made transparent what Schön (1983) has described as the "felt-path" through a building that an architect explores through his or her drawings, thus anticipating the experience of moving along a path.

Interestingly, the series of images that were chosen for publication give one the impression that they are a selection taken out of a continuum, like frames out of an animated film. In our view, the cinematic association is not an accident. When asked by Enrico Morteo in an interview (Stirling 1992), "Have your sources of inspiration always come from architecture or are there other influences?" (ibid., p. 19), Stirling replied, "They are mainly architectural. Perhaps the cinema may have been important to a certain degree. My generation grew up with the cinema . . . If I could, I went two or three times a week; I was obsessed by it and so it must have had an influence on me" (ibid., p. 19). The cinema is, of course, the ultimate medium for pictorial narrative and it is therefore quite plausible that Stirling, and others who followed suit, were consciously or unconsciously influenced by it. Cinematic imagery may well be one of the sources of what might be seen as "serial representation," which Stirling was a pioneer of, and which was later practiced by other postmodern architects like Rossi, Tschumi, Libeskind and others.

Figure and ground

Stirling's typical mode of designing, whereby he combined volumes (containing functional space) alongside primary and secondary axes of movement, and in the case of the museums also a variety of open spaces of different kind and size, yielded an intricate pattern of "solids" and "voids." Because of the primacy of circulation in Stirling's work the "voids" attain great importance, practically equal to that of the "solids" and not subservient to them, as is often the case in traditional architecture. To convey this state of affairs in drawings is not an easy task. Stirling had two main strategies to do so (not necessarily consciously): selectivity in representation and the manipulation of figure-ground relationships. The selectivity principle licensed the architect to present abstractions, selected elements: sometimes only open spaces, or voids, were drawn (e.g., Figures 2.6 and 2.7). But when looking at a drawing like this, which may contain nothing but a depiction of enclosed empty spaces, the voids become solids in terms of drawing conventions because they are the "figures" of the drawing.

The relationship between figure and ground is crucial to human perception and the interpretation of visual displays. Gestalt psychologists demonstrated that a set of simple principles governs the behaviour of the perceptual system and determines how we "see" compound configurations. In particular, what we understand to be a figure or a background is a function of the representation created by the perceptual system. In the words of Arnheim (1969): "All early imagery relies on the simple distinction between figure and ground: an object, defined and more or less structured, is set off against a separate ground, which is boundless, shapeless, homogeneous, secondary in importance, and often entirely ignored" (ibid., p. 284).

In some cases, however, a configuration is balanced such that figure and ground are reversible, as neither is perceptually stable unless viewed under certain constraints. The phenomenon of reversible figures was known since Necker reported it in 1832, following an experiment with an alternating perception of a line drawing of a cube that was consequently named after him the Necker Cube. The Gestaltists were the first to include this phenomenon in an overall theory of perception that was widely accepted and manifestations of it can be found in various fields, including the arts (particularly painting, drawing and engraving, e.g., work by Dali, Escher, and Alber). According to their account unstable perception, which leads to reversibility, occurs when conflicting principles of perceptual organization, determining meaning, are enacted. Later experimental data confirmed these theories (e.g., Chambers and Reisberg 1992). In the 1970s the newly founded cognitive science expressed renewed interest in reversible figures and the perception of figure and ground. This interest found its way into popular science and general public awareness (e.g., Attneave 1971) and possibly could have affected Stirling's choice of representational means. A case in point is Figure 2.9, where the opposing graphic means (shades) chosen to represent two levels of the Staatsgalerie in Stuttgart make it very hard to tell whether the central rotunda is meant to be the figure or the ground. Colquhoun (1984), who sensed the ambiguity that is so typical of reversible figures, wrote about the rotunda: "the geometrical center of the building has become a kind of negation – an absence rather than a presence" (ibid., p. 20). As a result the prominent role of open spaces devoted to circulation, given an equal status in the pictorial representation, was loudly and clearly conveyed to the viewer.

Collage

We stressed the fact that the museum publications presented images in the form of collages consisting of freehand sketches depicting anything from the entire scheme to a construction detail, hard-line axonometric drawings, of which many were partial and largely abstract, and, somewhat apart, also plans, and a few elevations and photographs of models. *Lotus International* (see note 3) added drawings from other projects and also a humorous drawing of Stirling sitting in his famous Thomas Hope armchair. This addition set the current designs in the context of the rest of Stirling's work. Why did Stirling choose collage as a means to represent his designs? Why the mixture of so

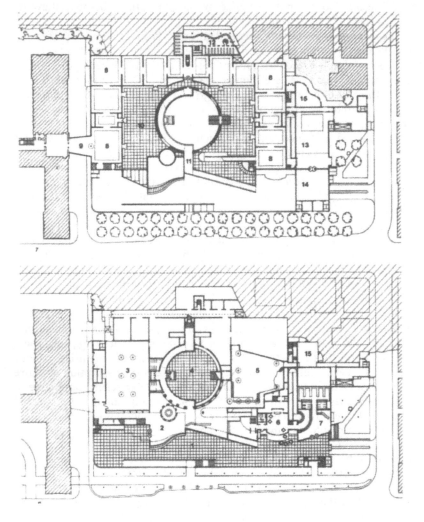

Figure 2.9 Two floor plans of the Staatsgalerie in Stuttgart.

many drawing modes and representational conventions, in a manner that at times appears disorderly and loose? No reference to an explication by Stirling himself has been found in the literature. We shall therefore offer our own conjecture regarding this question.

By offering many images, as is forcibly the case in a collage, it was possible to avoid a deterministic, final set of images. The collage – this particular collage – was therefore a way to stress the story, the narrative, the process related to the project, rather than a still and frozen end product. By opting to show preliminary sketches, including alternative design configurations, Stirling could remain somewhat ambiguous and non-committal, and he could advertise his taste for non-monumentality and eclecticism. Stirling's biographer says about him: "He liked those competition entries ['a house for Karl Friedrich Schinkel' - competition set by Stirling in Japan in 1979] which showed 'a delightful ambiguity . . .' He liked them, in fact, because they reflected what he had been doing at Stuttgart" (Girouard 1998, p. 207). Collage allowed Stirling to juxtapose urban scheme with detail, put conceptual problem next to ideas for technical solution, and expose precedent and its transformation. Along with figure-ground manipulation, collage allowed Stirling to maintain a certain level of desired ambiguity and, in Tzonis and Lefaivre's terms, also to respond to a timely "call for disorder."

The somewhat eclectic, fragmented display afforded by the collage appeared to reflect not only Stirling's ideological preferences, but also the actual nature of the designs in question. Stirling (1984) himself quoted Jencks' critique: "These drawings [Staatsgalerie in Stuttgart] accentuate the dualism inherent in the design, the juxtaposition of rectangle and circle, frontality and rotation, axiality and diagonality, and also the attitude of collage . . ." And in the words of Curtis (1984): "Collage seems to offer one of the central clues to the technique of the Staatsgalerie design. Throughout there are dramatic confrontations of images, forms, materials, themes. . . . Collage is a conceptual device, as well as a formal one, allowing ironical distance from the ethos behind past forms. It is therefore the ideal tool for the mannerist" (ibid., p. 42). Was Stirling a mannerist? To answer this question we must look at the three competitions together and contemplate similarities in the designs. Indeed, similarities are not hard to find, especially between the Düsseldorf plan and that of Stuttgart, which was seen by Stirling and his associates as "phase 2 Düsseldorf" (Girouard 1998). Both occupy a sloping terrain in a similar way, and share the idea of an open, central circular court. The similarity in programme and cultural context, the need to deal with urban issues and the appropriateness of pedestrian circulation as a generative concept, contributed to a close relationship between the projects. In addition, they were undertaken in close temporal proximity and Stirling admitted to a stylistic closeness between the designs: "The fact that our designs sometimes come in series has led me recently to believe that formal aspects may be stronger than I had thought" (ibid., p. 208) and "I cannot deny that there are stylistic similarities between buildings in a series, but they are worked out, perhaps exhausted, after three or four variants" (Stirling 1990, p. 13). According to Anderson (1984) many a good design is the result of continuity in design exploration across several projects. He sees this continuity as a system of research programs. A collaged presentation made it possible not just to represent a building or a scheme, but actually to illuminate the style, the formal aspects, through their many varied manifestations. It was a richer statement of design intentions, of what Stirling believed architecture is all about, than could have been achieved by mere factual accounts of the schemes.

We must also remember that Stirling was a lifelong friend and former student of Colin Rowe. Rowe was, of course, a co-author of *Collage City* (Rowe and Koetter 1978). Rowe's "collage" had a critical history background and motivation; it preached the value of variety and richness and advocated a mixture of styles – modern and traditional – allowing cities to grow out of multiple visions and artefacts of different nature and periods. It was a reaction to a certain sterility traced in the Modernist deed that was attributed to over-simplification. Rowe wanted complexity, which he equated with richness. He found it in Renaissance architecture and Stirling acquired that notion

Figure 2.10 A figure-ground plan of Wiesbaden, c. 1900 (Rowe and Koetter, 1978) Reproduced with the permission of MIT Press.

as his student in Liverpool (Wilson 1992). Despite some reservations, Rowe remained throughout his life an admirer of Stirling's work and wrote a lengthy introduction to Stirling's collected oeuvres (Arnell and Bickford 1984). He liked in Stirling the "magpie architect-*bricoleur*" (ibid., p. 16) who, because of these qualities, was well placed to promote compound and multifaceted architecture. In the Stuttgart project Rowe was charmed by what he saw as "an extensive series of episodes – entry sequence, ramp, stairs from court-yard to upper terrace" (ibid., p. 22). The traits of Stirling's buildings were congruent with Rowe's collage paradigm. Although *Collage City* is concerned with the built environment rather than with its representation in drawings, the book makes extensive use of collage-like techniques. These include full-page compositions comprising photographs, various kinds of drawings, diagrams and maps, and reproductions of artwork. Figure-ground reversals are very frequent, sometimes within the same plan, as illustrated by the plan of Wiesbaden that is reproduced in Figure 2.10. It seems that Rowe and Koetter selected a graphic mode of representation that came as close as possible to an embodiment of their profoundly postmodern arguments. According to them the book's text was completed in 1973 and, for the next few years, they collected and edited the illustrations which they considered crucial to the arguments they wished to present. Stirling, who designed and published the German museums while Rowe and Koetter were busy working on their book's graphics, acted in a similar manner: his buildings were composite masses, mini-cities consisting of "extensive series of episodes" that he wanted to represent in this way. The collage nature of the representations was designed to reflect the characteristics of the buildings.

Axonometric Drawings, Sketches, and the Design Process

As we already know, the axonometric drawings, and especially the "worm's-eye axos" that were included in the museum publications, were made after the fact in Stirling's office when the design work had already been completed. The architects named them "analytical drawings." The *Architectural Review* called them "after drawings" and later they were referred to in the literature also as "conceptual diagrams" or "schematic drawings" (see note 2). As mentioned above, the production of these drawings was not unusual in Stirling's office where they had become a standard routine, especially for the purpose of display and publication. Why were these axonometric views drawn? Certainly *not* in order to represent the way a building would appear to the eye, a task for which a perspective drawing is infinitely better suited. A partial explanation for the choice of axonometric down and, even more so, up views was given by Krier (see Girouard 1998), who testified that he had produced such drawings in order to project a particular image that he thought (and Stirling agreed) was especially appropriate to the nature of Stirling's work. Why project an image? Would the buildings and projects not speak for themselves if they were represented with faithful adherence to the way they actually appeared?

Evidently, this was not considered good enough. Stirling thought the appearance of buildings was not an independent aspect of their design. The appearance was a function of an overall composition of masses which were combined using functional criteria. Appearance was dictated by the choice of materials and by the syntax of joinery, at the level of detail and building-volume alike. Elevations were drawn late in the design process (Tzonis and Lefaivre's (1992) principle of "skin rigourism" may be helpful in explaining why a postmodern architect found it unnecessary to develop elevations early on). Indeed, aesthetic qualities were an outcome of design decisions made largely for other reasons. Therefore, there was no need for and no point in showing buildings as they appeared, as appearance is presented almost exclusively for the purpose of aesthetic demonstration and assessment. Were there any other reasons for representing buildings publicly? According to Stirling, there certainly were good reasons for doing so, but it was not so much the buildings themselves that he wanted to represent, but his thinking *about* the buildings that he wanted to expose as best he could. He was interested in emphasizing what he found most important in the designs, the reasons for acting as he did and making the design decisions that were eventually made. The "essence of the idea" and the "architectural understanding of the building, as distinct from an impression of how it might look in reality" were the important representational concerns (Wilford 1996, p. 32). The "essence of the idea" was the "underlying elements of [movement] continuity" which could not be fully expressed through conventional drawings, and the architects looked for a mode of representation that would do justice to the "clarity and dramatization of pedestrian circulation" (Wilford 1994, p. 5).

Axonometric views were a very appropriate medium for this kind of communication with the public. Unlike perspectives, axos show masses and can give information about their volumetric properties. Geometric relationships can be conveyed, and joinery explicated. They have analytic qualities and are capable of presenting design principles and concepts. These were exactly the things that Stirling wanted to convey. In addition, axonometric rendering makes it easy to be selective about what is included in the presentation, and therefore abstraction was made possible by way of extracting and drawing only limited components of the scheme. Hence elements of the "promenade architecturale," for instance, could be treated as an independent system without as much as a hint of the rest of the building and its features (e.g., Figures 2.6 and 2.7). By presenting not one but several axonometric views, it was possible to dedicate each to one specific aspect, one dimension of the overall message. As Evans (1997) made clear, a single architectural drawing "is always partial, always more or less abstract" and "never gives, nor can give, a total picture of the project" (ibid., p. 199). This is doubly true when it is not the end-state of the project that one wants to represent, but the ideas behind it. Up-view axos stressed this abstraction even further, and this is why they were made, as if to say: "do not expect a simulation of what you might see when looking at the building or at conventional representations of it. Here we treat you to an account of our rationale for having designed it as we have."

This implicit statement was innovative. "We do not face you with a fait accompli," the presentation announced. "We invite you to go back in time and witness the emergence of this design, join us in our thinking about it. Do not wait for what the critics have to say – we are giving you the information at first hand, for you to interpret it yourselves and to judge it as you will." This was, in fact, an act of great faith in the audience who received credit for being able to follow and understand the architects' reasoning. Stirling was a great believer in "democratizing" architecture, in the sense that it should be understandable and accessible to all people and not just to "specialists." Submitting his thoughts about the design of the building, and not just the building itself, to public scrutiny was therefore an uncommon and courageous act. But Stirling went even further in the process of exposing his thoughts to the public (a public of peers for the most part, however) by sharing his preliminary little sketches, the "doodles" that recorded thought-in-progress (Figures 2.1 and 2.2) and not only "after" analyses.

The "front-edge" phase of designing in which architects converse with themselves (and possibly with their close associates) about design ideas and options through rapid freehand sketches is considered very private, and was certainly seen as an intimate affair in the 1970s. To open up this private chamber and invite the public to take a look bears evidence not only to an unusual confidence in one's work, but also to a deep cultural change, one that proclaimed that the end-product is not the only thing that should be open to discussion. Instead, we can and should pay close attention to the process that brings it into being. The ultimate way to do so is to expose one's informal sketches, almost as students do when they discuss their work in progress with their studio instructors at architecture school. Preliminary sketches record fragmented images, questions one asks, experiments with shapes and forms, tentative ideas for possible solutions to big and small problems, as well as explorations of precedents and alternative solutions, even random visual patterns and configurations. Sketches do not reflect the design process, they *are* the design process. Their inclusion in a formal presentation of a project is therefore clearly intended to exhibit this process and render it worthy of study and discussion. The editors of *The Architectural Review* welcomed this novel representational approach and wrote in their brief preface to Stirling's report: "The Düsseldorf material in particular was intended as a demonstration of process: it includes conceptual doodles, design drawings, photos of the model and some 'after' drawings (which as a single image try to convey the essence of a project" (ibid., p. 289; see note 2). It seems that Stirling and the editors of *The Architectural Review* were attuned to the subtleties of changing architectural values and sensitivities. In fact, it would not be an exaggeration to claim that Stirling participated in bringing about these changes.

Here we must stress again that these tendencies were not entirely intrinsic to architecture alone. The 1970s were the era of "Conceptual Art," when intentions were of greater significance than artefacts. Cognitive science legitimized rigorous research of mental processes in all fields and new methods were being devised to study problem solving, especially creative problem solving. Newell and Simon published their much acclaimed *Human Problem Solving* in 1972. Arnheim's *Visual Thinking* had appeared in 1969 and contributed to our awareness that graphic expressions are important manifestations of human thinking, and not just externally communicated records of thoughts. Developmental and cognitive psychologists became interested in the act of drawing; attention was paid to children's drawings as mirrors of their cognitive and intellectual development (e.g., Gardner 1980; Goodnow 1977). A decade later the first serious studies of sketching in designing began to appear (e.g., Fish and Scrivener 1990; Goldschmidt 1991; Herbert 1988) and they have since been occupying a growing share of design thinking research. Sketches, we should add, are almost always made in sequences, or series. The reason is not only that a single drawing is forever a partial representation, as Evans (1997) so cogently pointed out. More importantly, designers make series of sketches because they build up and inspect their ideas gradually, and this is a process of trial and error and of dialectic reasoning that proceeds in small steps. Accordingly, many sketching acts are required that normally are spread over several sketches, sometimes an impressively large number of them (Goldschmidt 1991). For the purpose of studying a designer's thinking at the cognitive level, a complete set of sketches is required, preferably accompanied by the designer's commentary (think-aloud exercises are sometimes conducted for this purpose). Needless to say, this was not Stirling's intention in publishing some of his sketches, which he subsequently discarded as he estimated that their mission had been accomplished and they were of no further use. For Stirling, sketches were the first of three interconnected layers of work on the museum projects that he wished to expose together: initial explorations, final plans, and post factum "after" analytic axonometrics.

Summary: Representation as Reconstructive Memory

The notion of design "reconstruction" was introduced by Porter (1988) who defined it as follows: "By 'reconstruction' we mean a plausible way in which the design or building could be explained on the basis of evidence that it itself presents" (ibid., p. 170). The reconstructive act, according to this view, is carried out "backward from the design." Porter's purpose in undertaking reconstruction was to investigate how sets of ideas concerning place and architectonic type emerge and undergo a process of mutual adjustment. An understanding of this process was believed to facilitate descriptions that could have beneficial implications for computational design assistance.

We borrow Porter's notion of design reconstruction and use it for a different end. We see reconstruction as an after-the-fact interpretative act assumed in order to solidify the representation of a work of design so that it best describes its ideas and qualities, and in which it should be deposited in memory. Unlike Porter, we do not guess the ideas that serve the reconstruction or infer them from a normative set of design drawings. Rather, we look at interpretations made by the designer himself, in which he reconstructs the memory of the story of his designs. We refer to the representations of Stirling's museum designs as submitted by him for publication as acts of reconstructive memory. We postulate that a need for the reconstruction of memory arises when a representation is expected to convey a complex message that goes beyond factual information about a design product. The reconstruction is a design in itself; it is the design of design-image. Typically in such cases, the normative representational mode, through orthogonal projections and possibly perspectives or three-dimensional models, cannot capture the sought-after image in its totality, so that additional explanatory means are required. Such means can be quite diverse; they may include other graphic means as well as text, animation and more.[9]

An acute need to broadcast new messages is often felt during eras of cultural shifts. The means of representation are crafted after the needs they are expected to fulfil and, when culturally based needs change, old means might become inadequate or even inappropriate. In the history of architecture we can point to several examples of such eras with consequent developments in representational modes. Orthogonal projections were invented in Italy during the Renaissance and were instrumental in the gradual separation between design and construction. The designer, now no longer necessarily the master builder, was geographically remote from the site of construction and needed effective means to document his design intentions and communicate them to the builders. More recently, the birth of the Modern Movement around the 1920s brought with it representational innovations, especially where architecture had a close relationship with art. The Constructivists in Russia, for example, started using techniques like collage, photography, and abstract composition (two- and three-dimensional) to express their architectural intensions. Similar developments could be detected in Germany – for instance, in work that came out of the Bauhaus. The use of axonometric drawing was also revived in that period (Klevitsky 1997). It is therefore not surprising that postmodernism, once its ideas started to spread among leading architects and architectural critics, also demanded representational modifications to transmit those ideas. Postmodernism had a multifaceted vision which combined historicism, richness of expression, eclecticism, freedom to treat the envelope of buildings independent of their functional aspects, and a taste for a certain amount of disorder (by comparison with the strict order imposed by Modernist design principles). Stirling's work of the mid-1970s is considered postmodern in spirit by many observers. Regardless of whether or not he saw himself as a postmodern architect, he evidently felt that he was doing something new and different (also as compared to his own former work) that would not be understood unless he employed the appropriate measures to explain it to the public.

When reconstructing the memory of his design pursuit pertaining to the German museums, it was important to Stirling that the design story should be told, which returns us to the design narrative and its expression as presented and analyzed in this discourse. Stirling did not invent new representational means, nor did he feel the need to use ex-architectural media, as we may arguably call some of the notations employed by other postmodern architects a little later on.[10] Stirling's choice of representational substance was straightforward and utilized the material his office routinely worked with, including the tiny doodles and the abstract up and down axonometric views. The novelty rested in the decision to give priority to concept over factual description, to process over product, and to use for this end all available resources, including the public display of preliminary "raw" material that is normally made for internal consumption only. A generation later the architectural community worldwide conceives of Stirling's German museums in terms of both the built Staatsgalerie in Stuttgart and the publications of the 1970s. One has an image of the designs that is strongly impacted by the memory that Stirling had reconstructed for us, which would have had a presence and been influential even if the Staatsgalerie had never been built at all. It seems safe to assume that, had Stirling lived to witness the long-lived success of his representational choices, he would have been pleased, of course, but our guess is that he would have already been searching for ways to reconstruct a new genre of architectural memory.

Acknowledgements

This chapter is based in part on ideas developed within the framework of Ekaterina Klevitsky's doctoral dissertation in the Faculty of Architecture and Town Planning at the Technion. The writing of the chapter was partially supported by a grant to Gabriela Goldschmidt from the Fund for the Promotion of Research at the Technion. The authors wish to thank Michael Wilford for the kind permission to reprint sketches by James Stirling and drawings produced by James Stirling, Michael Wilford and Associates Limited.

Notes

1. We use the name Stirling to refer to the architectural office of James Stirling and Michael Wilford, who was an equal partner in the firm, as well as other associates in the office.
2. "Stirling in Germany, the architects' report", *The Architectural Review*, CLX(957): 289–296, November 1976.
3. James Stirling and Partner, *Landesgalerie* Nordrhein-Westfalen in Düsseldorf, James Stirling and Partner with Werner Kreis, Robert Livesey, Russ Bevington, Ueli Schaad; the Wallraf-Richartz Museum in Cologne, *Lotus International* 15: 58–67 and 68–79 respectively, 1977.
4. *Architectural Design* 47, (9–10), 1977: xx–61.

5. We did not find the diagrams in publications of the 1980s. Tzonis and Lefaivre (1992) published schematic drawings for the State Gallery project in Stuttgart and called them "conceptual diagrams" (ibid., p. 126). In Wilford, Muirhead and Maxwell (1994) the drawings are called "schematic down and up views" (ibid., pp. 58, 62).
6. At present Stirling's remaining sketches are being collected and archived. This endeavour is still in progress (Wilford 2000).
7. Various sequences of sketches and abstract diagrams for the German projects were published in the 1970s and 1980s by *Architectural Design (AD)*, *Lotus*, *Architecture and Urbanism (A+U)*, and *The Architectural Review*. It is important to point out that it was not accidental that these particular magazines were the ones to herald a new approach to representation of architectural images. According to Nesbitt (1996), the magazines *AD*, *Lotus* (founded in 1963), and *A+U* (established in 1971) had some of the same architects on their editorial boards, were among the independent magazines and academic journals that shared a "response to the professional crisis in modern architecture" (ibid., p. 23), and promoted new cultural sensibilities.
8. In Schinkel's building, however, the circular space had a roof over it.
9. Our thinking in this respect is not unlike that of Frédéric Pousin (1995), who addresses the appearance of new forms of architectural representation. In his view, new representational practices come into being when a need arises to rewrite history. Accordingly, architectural representation is not only prescriptive but also descriptive, and in this capacity it is used to express architectural thought both as a medium of conceptualization and the expression of doctrinarian thought. Pousin exemplifies his arguments through an analysis of Leroy's 18th-century work on the ruins of Greek temples.
10. We refer to notations used by Libeskind, Hadid, Tschumi and others.

References

Anderson, S 1984. Architectural design as a system of research programmes. Design Studies 5(3):146–150.
Arnell P and T Bickford, eds. 1984. James Stirling: Buildings and projects. New York: Rizzoli.
Arnheim, R 1969. Visual thinking. Berkeley: University of California Press.
Attneave, F 1971. Multistability in perception. Scientific American 225:62–71.
Baker, J 1992a. James Stirling and the promenade architecturale. The Architectural Review 191:72–75.
Baker, J 1992b. Stuttgart promenade. The Architectural Review 191:76–78.
Bois, Y-A 1981. Metamorphosis of axonometry. Daidalos no.15:41–58.
Chambers, D and D Reisberg 1992. What an image depicts depends on what an image means. Cognitive Psychology 24:145–174.
Colquhoun, A 1984. Democratic monument. The Architectural Review 176(1054):19–22.
Curtis, WJR 1984. Virtuosity around a void. The Architectural Review 176(1054):41–44.
Evans, R 1997. The developed surface: An enquiry into the brief life of an eighteenth-century drawing technique. In Translation from drawing to building and other essays edited by R Evans. Cambridge, MA: MIT Press, pp 195–232.
Fish, J and S Scrivener 1990. Amplifying the mind's eye: Sketching and visual cognition. Leonardo 23:117–126.
Gardner, H 1980. Artful scribbles. New York: Basic Books.
Girouard, M 1998. Big Jim: The life and work of James Stirling. London: Pimlico.
Goldschmidt, G 1991. The dialectics of sketching, Creativity Research Journal 4(2):123–143.
—— 1992. Serial sketching: Visual problem solving in designing. Cybernetics and Systems (23):191–219.
Goodnow, J 1977. Children drawing. Cambridge, MA: Harvard University Press.
Gössel, P and G Leuthäuser 1991. 20th century architecture. Köln: Taschen.
Herbert, DM 1988. Study drawings in architectural design: Their properties as a graphic medium. Journal of Architectural Education 41(2):26–38.
Jacobus, J 1975. Introduction to James Stirling: Buildings & projects 1950–1974 by James Stirling. London: Thames and Hudson.
Jencks, C 1982. Current architecture. London: Academy Editions.
Jones, E 1992. "Edward Jones" in Jim and I. The Architectural Review 191(1150):68–69.
Klevitsky, E 1997. Three dimensional composition in architectural education in the 20's: The Bauhaus and Vkhutemas schools. MSc thesis. Haifa: Technion – Israel Institute of Technology.

Klotz, H 1988. The history of postmodern architecture. Cambridge, MA: MIT Press.

Lacy, B 1991. 100 contemporary architects: Drawings & sketches. New York: Abrams, Inc.

Le Corbusier 1935 Oeuvre complète. Vol. 1. Zurich: Editions Girsberger.

Livesey, R 1992. "Robert Livesey" in Jim and I, The Architectural Review 191(1150):69–70.

Mendini, A 1984. Colloquio con James Stirling. Domus no. 51:1.

Nesbitt, K 1996. Theorizing a new agenda for architecture: An anthology of architectural theory 1965–1995. New York: Princeton Architectural Press.

Newell, A and HA Simon 1972. Human problem solving. Englewood Cliffs, NJ: Prentice Hall.

Pousin, F 1995. L'architecture mise en scène: Essai sur la représentation du modèle Grec au xviiie siècle. Paris: Editions Arguments.

Porter, WL 1988. Notes on the inner logic of designing: Two thought experiments. Design Studies 9(3):169–180.

Rowe, C and F Koetter 1978. Collage city. Cambridge, MA: MIT Press.

Schön, DA 1983. The reflective practitioner. New York: Basic Books.

Stirling, J 1975. James Stirling: Buildings & projects 1950–1974. London: Thames and Hudson.

—— 1984. Lecture '81 [quote from Jencks, C. (1979) in Architectural Design 49(8/9)], in Architecture in an age of transition edited by D Lasdun. London: Heinemann, p 197.

—— 1990. Design philosophy and recent work. Architectural Design 60(5/6):7–13.

—— 1992. Architecture in an age of transition (interviews with Enrico Morteo). Domus no. 741:17–29.

Tzonis, A and L Lefaivre 1992. Architecture in Europe since 1968. New York: Rizzoli.

Wilford, M 1994. Introduction to James Stirling, Michael Wilford and Associates: Buildings & projects 1975–1992 by M Wilford, T Muirhead, and R Maxwell. London: Thames and Hudson.

Wilford, M 1996. Our kind of architecture. Architects' Journal 203(23):32–33.

—— 2000. Unpublished letter to authors, May 12.

Wilson, C St John 1992. James Stirling: In memoriam, The Architectural Review 191(1150):18–20.

Designers' Objects

William L. Porter

Introduction

Buildings, especially great buildings, represent religions, nations and ideas. They embody and symbolize; as cultural artefacts they yield insights into that culture. Great buildings embody styles, traditions, types of buildings, ways of building, and even specific architects or firms. Buildings also evoke feelings and ideas that build on each person's experience, knowledge and skills, serving as a means to externalize, to reify an individual's own sense of the world and identity in the world. Buildings tell stories about themselves, their materials, their methods of assembly, their relationship to their locale; and they provide opportunities for contemplation, enjoyment, and inspiration that is independent of cultural or personal reference.

For some time I have puzzled over how, through their own processes of design, architects and designers may enhance the possibilities of creating meaning in all these different senses. In this chapter I have chosen to look at the things they make along the way of design. These things include drawings and other artefacts, both physical and digital. I have chosen to call them "objects" in order to emphasize seeing them in their own right, as well as in their more restricted role in approximating the final artefact.

What, then, is the relation between the objects the designer makes along the way of design and the last or ultimate object that results? The brief answer is intimate and independent. The relation is intimate in the sense that the objects are all made within the framework of a singular quest, that they may share similar properties, and may resemble one another. The relation is intimate in that objects along the way are intermediate with respect to the ultimate aim of the quest and are, therefore, a part of the evolving nature of the final artefact. The relation is independent in the sense that each object is made out of materials and methods of assembly that are specific to it, along with the associated crafts and skills. It stands to be examined in its own right for its own implications and meanings, regardless of what the designer is trying to achieve ultimately.

In their independence the objects along the way of design can assume the full significance of any object. They can embody, symbolize, and mean in ways that are identical to the cultural artefacts we identify as buildings or paintings or other "finished" works. They can be as carefully made, as beautifully conceived. In their fullness as objects in their own right, they serve to explore and express ideas that may or may not resemble or denote the ultimate design, ideas that nevertheless may play an important role in the creative process.

The interior designer's collage of materials and colours on a board has nothing to do directly with the geometry or functional layout of the space; it is meant both to convey the particulars of the finished surfaces and also to evoke the mood and character of the space and the resulting multiplicity of associations that can result. Another object along the way of design might be a narrative portrayal of an hour in the life of a proposed place of work, emphasizing the experience of those who work there, as well as the place-specific particulars that support and enable that experience.

Objects Made During Design

I see architectural production consisting of a series of objects. They mediate between the earliest thought and the finished building. Conventionally seen, the primary role of these objects is one of representation. But this is a restricted type of "representation." They are intended, from rough and early ideas, to represent the finished building with increasing accuracy. Their type varies between resemblance and specification. The accuracy may be visual, in creating a view of the building that will be identical with that view when the building is finished; or the accuracy may be dimensional in order that the contractor may know exactly the locations and sizes of the elements.

In contrast with their role in this type of representation, the making of objects for exploration is less widely appreciated. For designers, it can open up areas of creative potential in ways that the first role may not. In this chapter I stress how integral such objects are to the process of design enquiry and, in particular, how they enhance the architect's quest for meaning. Thus objects may be created that are not integral to the production of the building, yet are integral to the cultivation of ideas that relate to the building. It is the expressive content of these objects, as well as their representational link to the building, that makes them valuable to the designer as well as to others with whom the designer must communicate.

Exploration may be personal, to reinforce one's sense of the place, for example, to relate one's own size dimensionally to the place, to sense more vividly the geometry of a place through bodily actions and gestures, to sense the texture and feel of particular materials through touch, sight, and smell. It may be to study the building to see how it is made, to detect or infer certain kinds of assemblies or structural strategies. It may be to clarify semantic relationships to things and ideas that lie outside architecture, meanings that have accumulated from the past or formal similarities that give rise to unexpected associations. And it may be to address typological and conventional issues that are thought of as lying within the proper domain of architecture itself – for example, the degree to which a design tends to resemble specific building types – or to challenge conventional architectural understandings.

Designers employ objects to communicate ideas and information about a building in ways that a building itself cannot or may be less effective in doing. Sketches made after a building is built to convey its underlying concepts illustrate this idea. (For an example of this sort of sketch see Goldschmidt and Klevitsky, Figure 2.7.) The objects made after the fact of the design are different from the objects made along the way of design in that their reference is to something already made. In relationship to an intended or a finished design, the purpose of making objects can be to resemble and anticipate, to make more vivid certain kinds of experiences, to explore aspects in great detail, or to make evident specific ideas for the purposes of communication. But while their relationships to a design are of primary importance, their power rests in part on the strength of the design of the object itself, in its economy of means, in its elegance of line, in the degree to which it expresses and exploits the materials out of which it is made.

These attributes may, of course, be attributes desired of the final artefact, but achieving them in the object does not necessarily make that object resemble the final artefact, either in its form or its materials. Indeed, because these attributes depend on solving the design problem of the specific object, they will necessarily take a different form than if the same attributes were achieved in the final artefact.

The expressive power of the object depends on the excellence of its craft and design. As a result, it will be different from the ultimate design, if only because of differences of materials, methods of making, and scale. If well done, the object can exhibit intended qualities (of a high level of craftsmanship, handling of materials in ways that are appropriate to those materials, sensitivity to how the viewer will perceive the object given its size, level of detail etc.). Thus while a designer's object may fall short of exhibiting formal or spatial attributes of the ultimate object, it can, nevertheless, convey important ideas about that object.

It is in this sense that the final artefact is only another designer's object, subject to the same requirements of communicative power and the same discipline of conception, formation and execution of a particular set of materials. Of course, the final artefact can be a very expensive object, one that cannot easily be made out of its materials and at its scale many times over in order to improve it. If it were no more expensive than the objects made along the way of design, then the point about its being merely another object would simply be more obvious. Because it is so different in scale, in materials and in labour, the ultimate object must, therefore, be anticipated in ways that are different from how eventually it will be. And these differences obscure the essential similarities among the entire family of design objects associated with a particular design project, of which the ultimate object is a member.

Objects along the way of seeing

Designers may make objects for the purpose of understanding other objects. Travel sketches are an obvious example. Louis Kahn, Le Corbusier, and many other great architects have sketched and written in an effort to understand. In the balance between representation and exploration, the scales are sometimes tilted toward exploration. Examples are the columns at Karnak by Louis Kahn or the indigenous pottery from the Balkans by Le Corbusier, shown in Figure 3.1.

These sketches reflect a way of thinking about and experiencing works; or, put somewhat differently, they reflect the repertory of experience, thought and memory of the author in how those works are reconstructed in the form of the new object. They are what the authors make of the world, yet they are at the same time what the world makes of the authors – to anthropomorphize for a moment. The columns at Karnak *are* enormous and very closely spaced; the light and shadow fragment the columnar forms. And the pottery in its fascinating clutter of shapes and scales does, through "the magic of geometry [create] an astounding union of fundamental instincts and of those susceptible to more abstract speculation" (Le Corbusier 1989, p. 16).

It is different, to be sure, to sketch from an existing object rather than from an imagined one and, even more different, to sketch from a preliminary idea leading towards an idea not yet formed. Yet these activities have much in common: they are all efforts to get at some aspects of the idea in question, whether it is the profile, the quality of light, a way of utilizing materials, or an aspect of feeling. They all employ the same skills of the individual and they draw on the same repertoire of experience, images, and emotions, the same powers of observation, the same capacities to remember, the same penchants for (or against) abstraction.

For John Dewey "creative seeing is always a reconstructive act . . . For to perceive, a beholder must create his own experience – comparable to what the original producer underwent . . . without an act of recreation the object is not perceived as a work of art" (Dewey 1934, p. 54). Thus creative seeing and creative design are intimately linked.

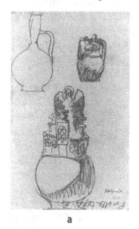 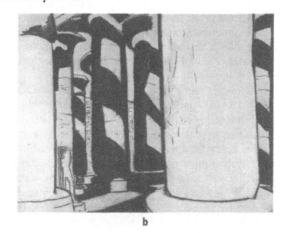

a   b

Figure 3.1 Travel sketches that creatively reconstruct. **a** Indigenous Pottery from the Balkans (from Le Corbusier 1989, p. 17, © 2003 Artists Rights Society (ARS), New York/ADAAGP, Paris/FLC). **b** Columns at Karnak (from Kahn 1962, Plate 10. Reproduced with the permission of R. Wurman).

On making things

As Goodman (1978) explains, people are "world-makers." They spend a great deal of their time carving out the niches, symbolic or real, in which they move about, so that they fit their aspirations and enhance their possibilities. They build cities and houses, but also musical compositions, poems and paintings. Young and old, people give form or expression to their ideas. They project their feelings outwards on to things, which in turn enables them to start a dialogue of greater intimacy and deeper understanding.

People are also "world-readers." They engage designed artefacts by actively reconstructing them through the lenses of their interests and experience. Bordwell (1985, p. 32), in discussing film audiences, argues that "The artwork sets limits on what the spectator does" but within these limits, the viewer literally recasts the play. What is true for a film audience is even more true for active explorers of interactive media, including games, learning environments, and good buildings.

People do not project on to neutral material. The world outside has something to say about how it can be constructed or how it can be read. Objects "afford" certain uses, practical or poetic, implying a nature that cannot be denied, regardless of how different the experiences of the user may be. And a fully developed design or work of art commands a certain interpretation, while still leaving open much for individual differences. The architectural artefact is possible to enjoy even by those who do not share and who cannot re-create the exact same experience.

Nor is the activity of projection itself devoid of its defining characteristics. Independent of the materials of the world, there are possibilities and limits to both perception and cognition that shape how we see and how we imagine. There are limits to the amount and complexity of material to which we can attend at any moment; our perception generalizes, groups, and performs other involuntary functions. Closer to our everyday world, there are limits to the number and types of things we can recognize, a limited number of metaphors that help us to organize and structure our experience, and, within our general propensity to group and see similarities, a heightened awareness of regularities and violations thereof.

Finally, there is a fundamental drive towards communication that influences most of us to seek those ways of seeing and imagining that others also may be able to see and imagine. Geometric ordering of a painting can be extremely subtle to the point of invisibility, although it may be rigorous and clever. But the paintings that communicate powerfully, through colour, shape, texture and the variety of means proper to the medium, both appeal to our sense of ordering and trigger our responses through cues that are specifically intended for that purpose.

These five kinds of actions – making, reading, engaging materials, projecting, and communicating – are as true of people in everyday life as of designers designing. Understanding them and appreciating their role in the design process can only help the designer to become more sensitive and insightful as both designer and researcher and more confident that he or she can reach a wider public.

Kinds of objects

Without creating a series of objects, the activity of design would be badly hampered. Such a series is particularly true of architecture and other design disciplines where the artefact to be designed is radically larger in size and employs materials and methods of construction different from those that can be used during the design process. In these design disciplines, therefore, the objects are not likely to resemble the ultimate object in all of its characteristics. Resemblance and representation in these objects are more distant from one another than an early sketch from an ultimate engraving.

Designers' objects can take a variety of forms. They can be text as well as drawn; they can be three-dimensional objects; or they can be narratives, taking the form of stories in graphic or written form, or in human performance. They can be blandly descriptive or richly evocative. They are in some sense like any other utterance or communicative product, but it is their content and purpose that distinguish them, not the medium of their expression nor even the particular form into which the medium is shaped. For example, a model of an existing building does not represent a design, but a similar model of a yet-to-be-constructed building does. One is intermediate in a process of design, the other is not. The interaction with the model in each case is different, as are the settings in which it is used, the people who will use it, and the purposes of its use.

Designers' Sense-Making

For the last few years I have given a course entitled Introduction to Design Enquiry for graduate students who already have, as a rule, their first professional degree in architecture. Typically, they are aspiring toward more specialized roles in practice, in teaching, and in research. It is the first course in a graduate programme that has been entitled Design Technology and concerns itself with such issues as the theory and practice of design, the building of tools to assist designers, the changing nature of the architectural artefact in society, and, of course, how we see, learn, behave, and design. Through this course I have both learned about and made use of the idea of designers' objects. I owe a large debt of gratitude to Dr Edith Ackermann who has developed and taught this course with me. She has brought an understanding of behaviour and development, as well as extensive experience in researching how children learn (see, for example, Ackermann 1996). We have become increasingly convinced that design can usefully be thought of as a way of looking at and acting on the everyday world, as well as a specialized activity associated with artefacts of various kinds and proper only to the design professions. Thus, instead of utilizing activities that are specifically associated with what designers are thought to do, I have taught it in part through activities that are part of everyday life and functioning. I have found that design informs the ways in which we ordinarily perceive and act. It does this by highlighting plausibility, cause, origins, histories, and associations. Design invokes personal experience and skills, and is driven by such things as curiosity, irritation, and a wish to complete.

The distance and the connections between perceiving and conjecture create a territory in which can be built the structures of communication between designers and those who use and experience designed objects. And if one accepts the proposition that design is integral to everyone's experience, then that same territory creates the potential for communication among all of us. I will illustrate some of the range of designers' objects by showing what students have done in a few of the assignments of the course.

Objects of objects

This assignment initiated our quest to understand how we engage worlds through objects and how we engage objects through worlds. We asked the students to find an object to reflect on and to make an object (a designers' object) that expressed their collective understanding of the object they found. We also asked that they keep track of the process by which they achieved that understanding.

A bottle of wine

In addition to their photograph and the text in the caption (Figure 3.2.) there was the important classroom "performance" of the three students who prepared the assignment. In this performance, they brought several different kinds of bottles (some pictured in the photograph), and they also brought glasses to illustrate the proper style of drinking the wine and, by implication, the sorts of circumstances in which the wine would be drunk. Taking all three "objects" into account – the text, the photograph, and the performance – they contextualized the object. Indeed, instead of bringing a single bottle, they brought several, already indicating the degree to which a "bottle of wine" was

Figure 3.2 Bottles of wine. If one tries to engage worlds through objects and understand how these objects actually enable us to understand the world, a bottle could reflect on this idea. We found this engagement in a "simple" bottle of wine for its functional aspects on one hand, as well as its semiotic aspects on the other.

The functional aspects of the bottle go far beyond a simple container of liquid. The particular shape of the bottle defines the quality of the wine, the position in which it should be stored and how it should be poured.

In addition to its functional aspects, the bottle communicates within itself the idea of gathering, tradition and culture. Thus a bottle gathers people together and becomes a vehicle to create an atmosphere. It forces itself from the outside into its own circle and everyone relates to this circle. Outside itself it has a meaning and within itself it has a meaning. From the outside it is a container and within itself it involves all of the senses of the body. All of these senses are utilized during the ritual of holding the glass and drinking the wine – sight, smell, touch, taste. Both the sensual perception and the intellectual perception are engaged. (Rita Saad, Maria Alexandra Sinisterra and Jennifer C.K. Seely[1]) an instance of a class of objects that had considerable currency. The text expands on this idea, helped already by dealing with a group of bottles rather than just one. The performance draws on their own experience and expertise, each giving a slightly different life context within which to see what the wine bottle is and what it means, and creating a narrative account through which it becomes more possible to understand the nature of the wine bottle.

Candle + glass

What were the designers' objects? The text, the reading of the text, the presentation of the merged objects in an atmospheric video and the objects themselves with a real flame on the table in front of us all. It is the combination of the candle and the glass and their new-found association into an object that resonates with a wide variety of human experience, not just those of the student team itself. It is even more provocative because of the unlikely role of the glass as candle-holder and, thus, the surprising new lantern (Figure 3.3). The performance in this case was more weighted toward the object in action – in flame – and toward the atmospheric video of the object in action, rather than the interaction of object-props, as in the last example, with the individuals of the team. The narrative constructions of the rest of us could be more personal, as we thought of such places and times of solitude, distant view, and impending darkness.

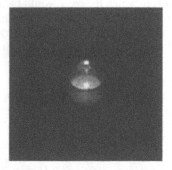

Figure 3.3 Candle + glass. We propose to explore a combination of two objects perceived by us as one – a plain candle merged with its holder. What is this object, what does it do for us, humans, and how does it affect the surrounding environment?

It is a minimal three-dimensional entity in the form of an inverted truncated cone made out of glass which lets the photons penetrate its body when the candle, usually made out of wax, is burning. Thus, it gives us light in a moment or a space full of darkness. The colour tones of the space around the candle are affected to reflect the redness of the holder. It gives us heat in a cold or hostile environment. It is also a generator of various smells. All these are purely physical qualities that are easily identifiable.

There is a much broader side to this simple object. It changes its appearance and meanings dramatically when it is affected by the flame. It suggests control as projected light can be manipulated; and lack of control because fire cannot be grabbed or shaped – it moves freely. And its usually pleasant smell and the silence of the burning flame can often touch on some long-forgotten memories. This makes it to some an originator of thoughts or even inspirations.

Its proportions, colour, material, and the effect on the environment, all combined, may be perceived by some as an entity with aesthetic qualities, resulting in feelings of pleasure derived from seeing beauty. Or it can also suggest time, human origins, adaptation and evolution. (Goncalo Ducla Soares, Alejandro Zulas, Victor Gane[1])

Narrative objects

Narrative objects are among the most important we make. They tie directly to our own experience through time; they depict episodes that seem to cohere; they suggest cause and effect; and at the same time they help us to see through others' eyes. They are in some ways ideal objects to start with, as they allow an intimate connection to be made between the self and the other.

Infinite Corridor

This assignment asks the students to observe the "Infinite Corridor," the main passage from the principal entrance of MIT to the interior of the academic campus: first to see it through their own eyes and then through the eyes of at least two authors. Thus there are objects: the essays that contain the students' observations and that reflect their experience and views; and the essays they write through the eyes of other authors that contain those authors' experience and views. The students were asked how one becomes confident that one's own knowledge is true of the world outside oneself ? And isn't this problem central not only to perception, but to communication with others? The caption of Figure 3.4 contains excerpts from one student's work.

Replicating objects

The exercise of "replication" has to do with "making" as a means of "reading" things. It is an attempt to understand the artefact through an exercise in creating plausible histories of how it might have come into being. This exercise has the advantage of reinvention where the invention is known. But the force of the artefact, its rationale, its raison d'être, can be understood through this device of reconstruction in ways that may otherwise be inaccessible. Through replication we may better be able to understand not only the characteristics of the thing, but also the world within which it was conceived and intended to function. In this assignment, we ask students to find a designed object of any size that they will replicate.

The Carpenter Center

In a series of drawings and photographs (Figure 3.5) this team emphasized as driving forces for the design the emphasis on the celebration of movement, allusions to highway ramps, the diagonal movement in Harvard Yard, the deliberate denial of façade, orthogonality, and frontality as indicators of the building's subservience to movement. This building by Le Corbusier was one of the most controversial ever built in Cambridge, seemingly out of place, and, for many, incongruous. However, what might look at first glance completely out of place turns out to have plausible design roots, both reinforcing and contradicting within the locale, as well as references to (very American) elements outside the locale – movement and the highway. Together, these elements make the Carpenter Center even more integral to its locale, highlighting things it chose to build on, as well as those things it chose to ignore or deny.

a

b

c

Figure 3.4 Three Ways of Seeing. **a** What do I see? An artery. Diversity. History. Fluctuation. Schedule . . . coupled with images of the classical columns at the entrance, the steps, a clock and other things. **b** What does Jacobs (1961) see? Does the corridor directly relate to a street or pavement? Is the Infinite Corridor MIT's most vital organ? Is the corridor safe? How much opportunity is there for crime in or around the corridor? Is there an unconscious, casual surveillance occurring in the corridor at different times of the day? Are there turfs present in or around the corridor? **c** What does Mitchell (1995) see? How many people wear headphones, talk on their mobile phones or email while walking down the corridor? Are we on camera, being projected into another space and time without our knowledge? How many signals of other people's interactions invisibly dance around us in the corridor? How long will it be before we blend into the architecture of the corridor? How long will it be before the walls of the corridor begin to be windows into other places and times? Is the corridor filled with a web of unseen space? Will the corridor ever physically quiet down because there will be no need for students to physically come to class? (Jennifer Seely[1])

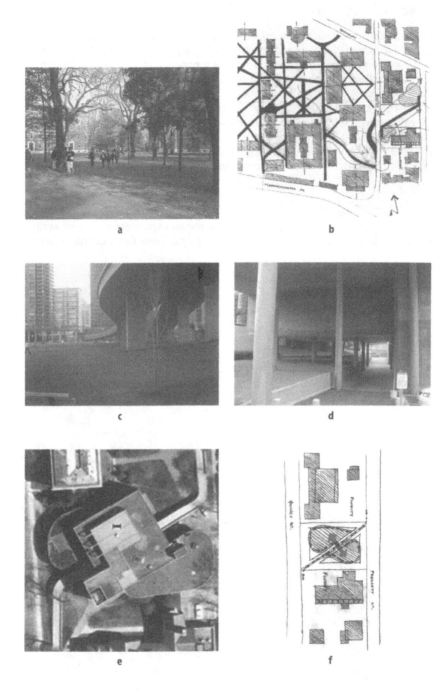

Figure 3.5 Carpenter Center Design Precedents. **a** Existing diagonal movement establishes a precedent for the- **b** Planned movement. **c** The highway interchange is a precedent for- **d** The structure of the Carpenter Center. **e** The building is shaped around- **f** A movement channel. (Alejandro Zulas, Goncalo Ducla Soares, Victor Gane[1])

Objects to read with

The aim of the exercise is to make sense of an American masterwork, the Exeter Library at Phillips Academy by Louis I. Kahn. The assignment calls for a four-stage process: first, the students' own personal experiences visiting the building, following their own interests, acknowledging and solving their own puzzles; second, the group's collaborative and collective efforts to produce an object that captures some important meanings of the building; third, their interaction with the class; and, fourth, their later reflections. It combines individual and group efforts in a synthetic approach to reading as a means to improve both individual and collective understanding. It also proves to be a vehicle for the improvement of communication.

Team 1

This team emphasized the geometric perfection and symmetries, as well as the doughnut-like layering of the building. They embodied these characteristics

Figure 3.6 Expression through craft. **a** Craft of material selection. **b** Sizing. **c** Assembly. **d** All to effect. (Rita Saad, Alexandra Sinistera, and Jennifer Seely[1])

of the building in their designed object through the craft of its assembly in the careful sizing and placement of elements, and its composition. That permitted the dematerializing results of the finished artefact – when illuminated from the inside, their object translated structure and surface into light and space, dematerializing its physical elements. This same translation also occurs in the actual building itself (Figure 3.6).

Team 2

This team captured the search for knowledge they felt the building embodied by choosing the metaphor of the book. Their book, like the building, has a plain and planar cover. Once inside, it reveals itself in increasing geometric complexity. Connection from one page to another is made evident through holes cut in each page. The second half of the book is symmetrically opposite to the first half, making it possible to read the book from either end, celebrating its symmetry and the book's indifference to the direction in which knowledge is sought (Figure 3.7).

Team 3

This team exploited unlikely juxtaposition to reveal plausible worlds in which the building might be imagined, as well as to emphasize the particular qualities that set the building apart from others (Figure 3.8). For example,

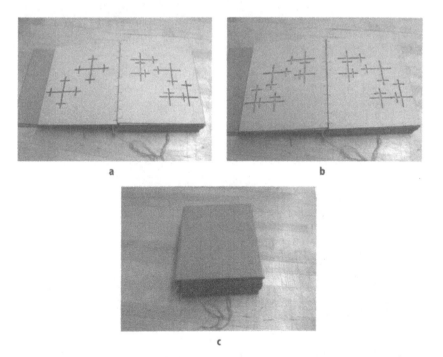

a

b

c

Figure 3.7 The metaphor of the book. **a** Increasing geometric complexity to **b**, from a deceivingly plain exterior, **c**. (Hans-Michael Foeldeak, Janet G. Fan, Konstantinos Tsakonas, and Luke Yeung[1])

replacing the concrete of the interior with brick forces thought about why, how and where materials are used, which, in this building, is rewarded by the discovery of remarkable consistency and expressiveness. And resiting the building in the city forces increased attention to its shape, its contour, and its relationship to its Georgian neighbours and its placid surroundings. By such juxtapositions with both likely and unlikely contexts, the team narrowed and deepened the conceptual region in which the building finds itself. They helped to articulate the different associations, as well as to reveal some of the particular backgrounds and experiences that each of the team members brought to the quest, permitting a shared space of discourse to emerge. This enlarged space was one into which, because of its diversity and evident humour, many others could enter.

Team 4

Team 4's approach was through a dominant metaphor of light which revealed in discoverable geometric complexity an inner glow that refers not only to the ineffability and desirability of knowledge, but also to the structure of light in

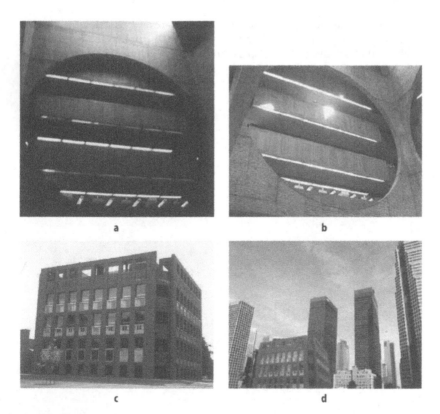

Figure 3.8 Unlikely juxtapositions. **a** Interior as it is, in concrete. **b** As it might be thought, in brick! **c** Exterior as it is, on a placid Georgian campus. **d** As it might be thought, in the city! (James Tichenor, Stylianos Dritsas, Keru Feng, Sameer Kashyap, Johanne Blaine and John Alex[1])

a   b

Figure 3.9 Source of light. The building as a set of screens, **a** mitigating the experience, **b** of a flickering interior (candle) light. (Victor Gane, Alejandro Zulas and Goncalo Soares[1])

the building itself with its inner court which is naturally yet mysteriously lit. Thus the new object pointed up the power of the building itself to express the same set of ideas (Figure 3.9).

What Propels Discourse with Objects?

The evident enthusiasm of the students in building these various objects masks a more important phenomenon: their increased security of understanding of the very things that these objects are intended to elucidate. I venture to say that their new understanding stands the test of their own subjective evaluation, as well as evaluation by others.

There is likely to be a high degree of integration of the object's anticipated perception and use with the experiential repertory of the designer who made it. Such objects can be closer to the designer's own experience, more directly subject to his or her own direct action and reaction, more a function of his or her own skills and predilections than buildings or other objects whose production depends on means outside the direct control of the designer. And these objects are made in an effort to understand, to bring one's own experience into relation with what is seen. At the same time, there is a strong impulse toward communication. This process of articulating the making sense of things (and the making of things that make sense) that can be scrutinized and read by others is not only to validate one's provisional understanding, but, as Reddy (1985) might argue, to create part of the network of shared ideas that constitute the fabric of society.

The set of things built during the processes of designing (or of reading objects already in the world) can be seen as components of a language by means of which those objects can be talked about. Following from Wittgenstein (1965, 77 ff.), language and behaviour are built in this way, generally in our culture and specifically within a work of art. Most artists are aware of the constructive engagement called for on the part of the viewer. I believe that the potential for the constructive engagement of the viewer with the ultimate object may be increased if the designer has gone through an eloquent history of its making. The investment in the making of things along the way of design can be great and varied. Emotionally, intellectually, bodily, aesthetically, each person comes not only to know but also to build the durable means of knowing, means that are inextricably associated with the objects that he or she has made. And because these means are contingent on location in particular places and on available materials and technologies, it is not only the subjective and inter-subjective power of this language of design that results, but also the cultural specificity as well. The designing of a work of art as contrasted with functioning in everyday life relates to the development of a language and the cultivation of behaviours that are specific to the work and that afford experience beyond the ordinary. This is a matter of degree, not difference.

Finally, what is the relation between designers' objects and design representation? These objects reach far beyond representation in the narrow sense of resemblance or specification. They can represent the qualities of the intended design or even the mood it is supposed to evoke. Designers' objects can reflect an investigation into the properties and behaviours of certain materials that represent a craft attitude in general or a way of working with particular kinds of materials that will be present in the ultimate design. They can aid in defining context, exploring alternative surroundings for a building or alternative materials of which it might be constructed, raising questions to be answered later in the design quest. They can be seen as integral to the process of learning that occurs during the design process, in which they surface ideas, elements, properties, and relationships that can become appreciated and later appropriated into the designer's stream of thought. Together these modes of representation create a setting within which the designer can achieve expressive intent. Designers' objects represent aspects of designers' worlds.

Notes

1. Introduction to Design Inquiry in the autumn of 2002 included the following students: John Alex, Johanne Blain, Sylianos Dritsas, Janet Fan, Keru Feng, Hans Michael Foeldeak, Victor Gane, Sameer Kashyap, Rita Saad, Jennifer C.K. Seely, Maria Alexandra Sinisterra, Goncalo D. Soares, James Tichenor, Konstantinos Tsakonas, Alejandro Zulas. Guests in the seminar included George Stiny, Terry Knight, William Mitchell, and Takehiko Nagakura, faculty; and visiting faculty Edith Ackermann and John Gero in the Department of Architecture. Eleanor Fawcett was the Teaching Assistant; and Paul Keel, Yanni Loukissas, Ben Loomis, Axel Kilian, and Janet Fan, advanced students in the program, were consultants. Figures and captions are used with the permission of the sudents.

References

Ackermann, E 1996. Perspective-taking and object construction: Two keys to learning. In Constructionism in practice: Designing, thinking and learning in a digital world, edited by Y Kafai and M. Resnick. Mahwah, NJ: Lawrence Erlbaum Associates, p 25–35.

Bordwell, D 1985. Narration and the fiction film. Madison, Wisconsin: University of Wisconsin Press.

Dewey, J 1934. Art as experience. New York: GP Putnam's Sons. Tenth printing 1958.

Goodman, N 1978. Ways of worldmaking. Indianapolis, Indiana: Hackett Publishing Co. Fifth printing 1988.

Jacobs, J 1961. Death and life of great American cities. New York: Random House.

Kahn, LI 1962. Drawings. In The notebooks and drawings of Louis I. Kahn, edited by R Wurman and A Feldman. Philadelphia: The Falcon Press.

Le Corbusier 1989. Journey to the east. Cambridge, MA: MIT Press.

Mitchell, WJ 1995. City of bits [electronic resource]: Space, place, and the Infobahn. Cambridge, Mass: MIT Press.

Reddy, MJ 1993. The conduit metaphor: A case of frame conflict in our language about language. In Metaphor and thought, edited by A Ortony. Cambridge: Cambridge University Press, p 164–201.

Wittgenstein, L 1965. The blue and brown books. New York: Harper Colophon edition.

PART II

From the Perspective of Engineering

4

Distributed Cognition in Engineering Design: Negotiating between Abstract and Material Representations

Margot Brereton

Introduction

Recent writings on cognition have focused on the contribution of external representations in supporting internal thought processes. External representations have been found to be so instrumental to thought that cognition is described as "distributed." "Distributed cognition" refers to the way that cognitive achievements arise from not only the internal thought processes of people, but also from the external representations such as the material systems, sketches and information technologies with which they work. The term "distributed" also refers to the fact that thinking processes may be distributed among members of a social group. This chapter focuses on how material representations – prototypes and bits of hardware – are instrumental to thinking in engineering design. In the closing discussion, extensions to other fields of design are considered.

In influential early writings on distributed cognition, Hutchins (1995) described the activities of navigation as the cooperative achievements of humans with technologies. Goodwin (1990) discussed interactive processes that involved technologies and physical apparatus on an oceanographic research vessel. Chaiklin and Lave (1993) analyzed how a blacksmith worked with materials as he designed and made a piece. In related design research, Schön (1994) described the process of designing as a reflective conversation with the materials of a design situation. Harrison and Minneman (1996) described how objects are an integral part of design communications, altering the dynamics in multi-designer settings and forming part of the pool of representations that are drawn on by designers.

While this chapter uses the term "distributed cognition" to describe the process of designing and developing design understandings, this should not be taken to imply a commitment to the cognitivist tradition of studying thinking. No attempt is made to understand or depict the internal workings of the designer's mind, to which we have no direct access. Further, this chapter does not speculate on the issues of perception or internal representation.

The epistemological stance taken here recognizes that the designer's internal thought processes contribute to the development of design understandings, but regards knowledge as fundamentally social in origin – that is, knowledge and information lie within the social milieu of people, artefacts, books, the world etc., and people access and construct this information into personal knowledge through their interaction with the social milieu and the lived world. This stance is drawn from the tradition of phenomenology after Husserl (Macann 1993), the extension of phenomenology to the social world after Schutz (1932; see also Dourish 2001) and the social constructivist learning philosophy attributed to Vygotsky (Moll 1990). Hence the study of design described in this chapter used methods that were derived from the social sciences.

This chapter demonstrates how material systems are used in engineering activity to support: a) thinking in design and b) learning fundamental engineering concepts. Specifically, I will demonstrate that learning in design is enabled through continually challenging abstract representations against material representations. This comparison between representations reveals gaps that inspire further design activity. The cycle of representation and re-representation in abstract and material forms advances the design, advances the designer's understanding of design requirements, reveals hitherto implicit design assumptions, and extends the designer's hardware repertoire of familiar hardware components and machines. In addition to informing design activity, this process of re-representation assists students in sorting out fundamental engineering concepts. The chapter draws on videotape data of design activity in order to illustrate the use of material and abstract representations by designers. First I define the different types of representations that are referred to below.

Types of Design Representations

There are four dimensions with which it is useful to classify representations when considering their use by designers. These dimensions are shown in Figure 4.1.

Internal vs. external

Representations can be classified as internal or external. Internal representations are those thoughts in the designer's mind to which a researcher does not have direct access. External representations of design thinking are spoken utterances, written lists, drawings, prototypes, etc. These external representations are directly available to the design researcher and to other designers during the normal social interactions of a design team.

Internal	$\longrightarrow$	External
Transient	$\longrightarrow$	Durable
Self-generated	$\longrightarrow$	Ready-made
Abstract	$\longrightarrow$	Concrete (Material)
	$\longrightarrow$	

Figure 4.1 Dimensions of design representations.

Transient vs. durable

Many design representations are transient, produced in the act of designing but never captured. Words articulated and gestures gesticulated in a design discussion are transient external representations. Similarly, transient information produced as a machine is being tested is often never captured. At the other end of the scale, durable representations are those sketches, drawings, and physical prototypes that endure and can be kept and referred to. They are often used as communication devices at meetings and they form the basis for further design developments.

Transient representations play a large part in shaping the design process and the final result. Brereton et al. (1996) demonstrated how words articulated in the negotiations of a design team drive the final product. Designers use negotiation strategies such as referring to third parties, standards and experience in order to promote their preferred design alternatives. Ideas that get discussed and rally support stand a greater chance of being developed into sketches and prototypes.

Self-generated vs. ready-made

Self-generated representations are produced by the designer in the act of designing, such as words articulated, sketches produced, and CAD (computer-aided design) drawings drawn. In addition to generating their own representations, designers often seek out ready-made pieces of hardware in their environment and gesticulate with them in order to help them think through an idea. Ready-made hardware is used because it has particular properties that assist thinking and it is readily available. In many cases a quick model is more important than an accurate model and prototyping or gesticulating with readily available pieces of hardware will get to a useful representation quickly. In the case of gesticulating with pieces of hardware, the representation has elements of ready-made, self-generated and transient. This fact is understandable when one recognizes that representations in design are continually evolving.

Abstract vs. concrete

Representations describe designs at various levels of abstraction. On the more abstract end of the scale lie lists of requirements, sketches, and scale models. A brief written list of requirements is abstract because, although it says what functions the design should fulfil, it does not specify the design. A number of different physical configurations could fulfil the same set of design requirements. Sketches are abstract because they leave much detail undefined.

As a result they can be interpreted in various ways. Sketches are often the preferred means of representation and communication at the idea generation stage, precisely because they do not force the designer to pay attention to details that the designer is not yet ready to consider (this is why it is so frustrating to use a CAD system when one wants to sketch). Issues of physical scale can also be considered on this abstract-concrete dimension. A scale model is usually more abstract than a life-size model. It usually has fewer details defined than the life-size model.

On the more concrete and specific end of the scale lie engineering drawings and physical machines. In contrast to a sketch, an engineering drawing of a design is very specific. Engineering drawing conventions were established to ensure that a trained machinist can interpret a drawing in only one way, so that he or she will build exactly what the designer intends. A drawing specifies a range of tolerances on each dimension and it specifies the materials to be used. However, an engineering drawing is still only a set of instructions to be interpreted. It has none of the three-dimensional, material properties of a real physical machine. Once a machine exists it is a unique piece of hardware, with unique dimensions. Its materials and form embody the history of all the manufacturing processes it has undergone and the cycles of loads to which it has been subjected. It wears uniquely, according to its context of use and its history of manufacture and operation.

Although representations vary from abstract to concrete, this scale alone cannot characterize the level and quality of information in a representation. Different representations make different kinds of information available. The way in which designers must interact with the representation in order to get the information that they need is an important factor for determining the usefulness of a representation. For example, a designer can rotate a shaft in its bearings in a physical machine in order to get a feel for the amount of play in the bearings. However, if the designer wants to know the nominal dimension between two shoulders of the shaft to within a millimetre, a measuring instrument must be obtained. In this case an engineering drawing, which is a less specific representation, would reveal this dimensional information directly through visual inspection. Just because a machine is more specific than a drawing does not mean that it conveys all the information about the machine to the designer more readily. Similarly, a computer simulation conveys different information and allows different design enquiries than a physical prototype. In particular, it is easier to adjust parameters and ask "what if" questions of a computer simulation. In contrast, a physical prototype is durable and cannot be changed easily; however, it can be used and manipulated in a context of use that allows the designer to understand human-machine and machine-environment interactions in a way that is difficult to do with even the most advanced computer models. Employing a physical prototype in a real context of use often reveals unanticipated information, which is one of the strengths of physical prototypes. So, the level of abstraction of a representation cannot characterize the kind of information that is available in a representation. The suitability of a representation to a task depends on the enquiry that is being undertaken by the designer.

It was tempting to include a scale that refers to whether a representation is "direct" or "indirect." The only "direct" representation of a design would be the final built machine. All other representations would be "indirect," alluding to what will be the final design. However, taking this approach would imply that the final design is the reference point for the design process. From the point of view of designing, the important reference point is the current understanding of the design, which is distributed among the members of the design team, and the activities that the designers need to undertake in order to advance the design towards the design requirements (recognizing that the evolution of these requirements is also part of the process of designing). Seen from this point of view, the notion of the "final design" seems relatively inconsequential. It could also be argued that designs are never finished. Hence, I decided that there was no need for a scale to compare the current state of the design with the "finished design." The abstract-concrete scale would suffice to represent issues that relate to the flexibility and directness of interpretation. The two key points about the abstract-concrete scale are that a) abstract representations offer more flexibility in interpretation than concrete ones, and b) different representations convey different kinds of information more directly than others. The suitability of any given representation to a task depends on the information sought by the designer.

Material representations

In this chapter I am concerned with material representations – pieces of hardware – and the way in which they are used in designing. Material representations are external representations. They are specific concrete physical representations. Because they can be reconfigured, they vary on the scale from transient to durable and self-generated to ready-made. The ability to reconfigure and reinterpret material representations is where their power lies in helping designers to think and learn. This chapter illustrates the way that material representations are challenged against abstract representations, such as design requirements, in order to advance the design. It illustrates the way that theoretical predictions derived from abstract fundamental concepts, such as Newton's laws, are challenged against the behaviour of physical devices in order to refine the theoretical model and further the student's understanding of engineering fundamentals. The latter is dealt with more thoroughly in Brereton 1998.

Hardware

I use the term "hardware" as shorthand to refer to material representations in general. The term covers raw materials such as string, cardboard, wood, and steel, as well as physical devices that have been fashioned from raw materials. (I regard sketches as abstract representations drawn on a piece of physical material, not material representations in and of themselves.) I use the term "physical device" when I am referring specifically to pieces of machinery and not to raw materials.

Data Gathering

Several engineering design activities were videotaped and analyzed in order to determine the roles played by hardware in supporting group activity and the resulting contributions to design thinking (see Figure 4.2). For a full description of these activities and the larger study from which this chapter is derived, see Brereton (1998). The activities varied from a concept sketching exercise in which no hardware was available to projects in which a kit of hardware was provided and augmented with other hardware as the designers saw fit. The activities were:

1. a conceptual design session in which groups of students sketched ideas for a kitchen scale mechanism (no hardware);
2. a design and build exercise in which kit hardware was used to build an aluminum crane (kit hardware);
3. a design project in which a group of students designed an energy efficient model All Terrain Vehicle using a Lego kit, a choice of motors and other hardware (kit hardware and evolving project hardware).

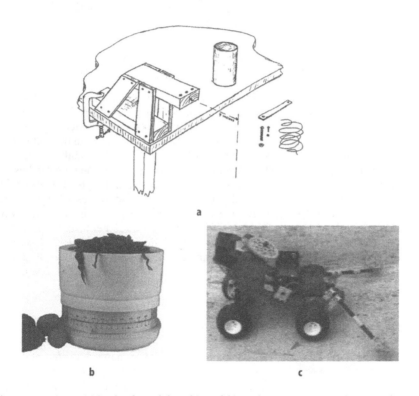

a

b   c

Figure 4.2 Design activities that formed the subject of this study.
a Crane design. Design a structure to support a 20lb load 9in out from the end of the platform using a kit of hardware that included aluminium strips, nuts and bolds, screws and string. Duration: 90 minutes. Exercise and sketch from Miller (1995).
b Kitchen scales mechanism conceptual design. Design an internal mechanism for the scale concept shown above. The mechanism should transfer the weight from the scale pan into the rotation of a pointer in the horizontal plane. Exercise duration: 30 minutes.
c Energy efficient all-terrain vehicle project. Design an efficient model all-terrain vehicle using Lego, and batteries and motor of your choice. This vehicle must cross a stretch of gravel and climb a carpeted ramp using as little energy as possible. Energy consumed during performance trials will be measured by instrumenting the vehicle to determine the average current and voltage levels. Project duration: two and a half weeks.

Analysis Method

Video Interaction Analysis (VIA) (Jordan and Henderson 1995) was used to investigate the dialogue and gestures of students as they worked with hardware. VIA is a qualitative analysis method that has its roots in the social sciences. It is suitable for helping to formulate hypotheses and find patterns in complex data. Video captures social interaction and learning activities as they occur and yet allows playback and scrutiny. It provides access to conversation, gestures, expressions, actions, and the immediate workplace context. It allows repeated viewing of the original data to examine the consistency and generality of the observations. It reveals the unanticipated and immerses us in the activity with the student.

In Video Interaction Analysis, the primary investigator watches each tape, making a log of the content and selecting segments of tape that are representative of the activity or of particular interest. An interdisciplinary team then observes the selected segments of tapes and identifies routine practices, problems, and resources for their solution. Only those practices confirmed by the raw data that occur repeatedly in different parts of the tape are admissible in the analysis. Conjecture that is not supported by the video data is dismissed. Activities do not reveal the individual cognitive processes of learning, but they reveal all the verbal and gestural interactions – that is, the inputs and outputs of the individual thinking processes that were made available to the group. Thus, they provide the researcher access to the external representations used in activity.

Some fundamental assumptions of the Interaction Analysis method are that:

1. Knowledge and action are fundamentally social in origin – knowledge and information lies within the social milieu of people, artefacts, books, the world etc., and people access and construct this information into personal knowledge through interaction with the social milieu.
2. Theories of knowledge and action should be grounded in verifiable observable empirical evidence.
3. Theorizing should be responsive to the phenomenon itself rather than to the characteristics of the representational systems that reconstruct it – analysis is done directly on videotape data, rather than on transcripts, or other reduced forms of data. It is acknowledged that video does not capture the broader context of events in the videotape and that the view from the camera is the only view available.

Several Video Interaction Analysis sessions were undertaken with people from a variety of different disciplines: engineers, architects, social scientists, cognitive scientists, education researchers, computer scientists, linguistics researchers, and anthropologists.

Results of Video Analysis

The video analysis of the activities revealed that hardware plays a very formative role in learning and designing, rather than simply serving as a final physical testing ground for ideas that have been developed through abstract reasoning. Hardware acts as an intermediary through which students develop and convey their thoughts to each other. Learning is constantly mediated through feedback from hardware or, in the absence of hardware, reasoning based on physical experience.

Negotiating between abstract and material representations

The general process of learning and designing with hardware is shown in Figure 4.3. The designers negotiated the demands of the requirements (abstract representations) against the performance of the current prototype (a material representation) and tried to bridge the gap by making design proposals, taking actions until the two reached a satisfactory agreement. This process led designers from the externally defined requirements and their own theoretical and hardware starting points through to a refined understanding of theoretical concepts and an extended hardware repertoire.

A plot of references to material in the workspace and to abstract constructs (such as design requirements, theories, and functions) in the crane and scales exercises revealed that the references were heavily interleaved, as shown in Figure 4.4. This quantitative analysis is described in Brereton and Leifer (1997) and Brereton (1998).

A series of events from the crane exercise that illustrates a typical process of designing with kit hardware is shown in Figure 4.5. The design advanced through students making and testing design proposals. These design proposals arose from seeing possible configurations of the ready-made kit hardware that would meet the design requirements. Design proposals were made through gestures with hardware augmented with speech (transient representations). Each proposal introduced a hardware configuration supported by rationale that referred to a physical property such as "strength" or "torsion"

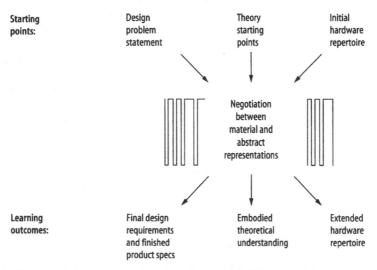

Figure 4.3 A negotiation process leads students from the task definition and their own theoretical and hardware starting points through to a refined understanding of theoretical concepts and an extended hardware repertoire.

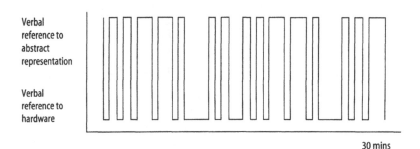

Figure 4.4 A plot of the crane design activity revealed that the discussion consists of interleaved references to abstract representations (design requirements or theoretical concepts) and hardware in the workspace.

(Figures 4.5a and b). Presumably, a design that would meet the requirements would need these properties, so these properties were used as a sort of shorthand way of referring to the requirements (support a 20lb load, 9in from the end of the platform). Once a prototype existed (Figures 4.5c and d), each design proposal sought to modify the hardware (shorten the diagonal strips, remove the brace) in order to change the properties (length, tendency to twist) that became apparent from the prototype behaviour. Design progressed through evaluating the hardware with respect to the requirements, making a proposal, implementing it in hardware, and then re-evaluating the hardware with respect to the requirements. This behaviour was typical in all groups that were studied designing cranes and in other kit design exercises.

The design process described above is negotiation in two respects. First, the designer argues the requirements (abstract representation) against the performance of the current hardware prototype (material representation) and tries to bridge the gap. Second, in group-work, the students negotiate one student's opinion against another and try to reach agreement among themselves about how to proceed. In the end, the hardware specification must satisfy the requirements, so there is a sense in which the requirements (abstract representation) and the hardware (material representation) must converge through the design activity; however, many divergent paths may be taken before final convergence is achieved.

Schön (1983) described the process of architecture students sketching as involving a reflective conversation with the materials of a design situation, the sketch talking back and revealing issues to the designer. The sketching process relies on the ability of the sketcher to interpret and modify the sketch, to see issues presented by the sketch. The evolving physical prototype is a yet more active and evocative participant than the sketch. It responds through physical behaviour. It may deform under loading, make noises, smell, wear, or jam. It is sensitive to attachment procedures. It is intolerant of poor assumptions or overlooked details that may not reveal themselves in a sketch. It reveals or suggests such oversights through its behaviour. The student gets feedback through seeing, feeling, smelling and hearing the prototype. As students become more acute observers, they learn to experiment and probe actively, watching, listening, touching, and smelling the prototype. They make

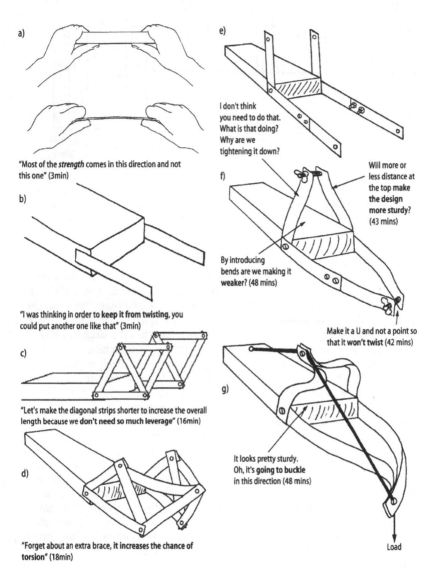

a)

"Most of the *strength* comes in this direction and not this one" (3min)

b)

"I was thinking in order to **keep it from twisting**, you could put another one like that" (3min)

c)

"Let's make the diagonal strips shorter to increase the overall length because we **don't need so much leverage**" (16min)

d)

"Forget about an extra brace, **it increases the chance of torsion**" (18min)

e)

I don't think you need to do that. What is that doing? Why are we tightening it down?

Will more or less distance at the top make **the design more sturdy?** (43 mins)

f)

By introducing bends are we making it **weaker?** (48 mins)

Make it a U and not a point so that it **won't twist** (42 mins)

g)

It looks pretty sturdy. Oh, it's **going to buckle** in this direction (48 mins)

Load

Figure 4.5 Students pit abstract requirements against hardware behaviour as they design the crane.

causal links between actions and behaviours. To learn from a hardware prototype, the designer must interpret the physical response and decide on the next move.

The Roles of Hardware in Learning

So far I have identified a learning process that is almost identical to Schön's (1994) notion of a reflective conversation with the materials. Schön (1990) described the design process as consisting of the framing of the design problem, the discovery mediated by the materials, and the subsequent reframing of the problem in the light of the discoveries made during designing. The contribution of this work is to identify just how the materials – in this case hardware – mediate the learning process and to identify specifically what kinds of things are learned through use of hardware to support thinking. Hardware has been identified as playing the following roles in mediating the learning process, as described in Table 4.1:

1. Hardware as a starting point.
2. Hardware as a kinaesthetic memory trigger.
3. Hardware as a thinking prop.
4. Hardware as a chameleon.
5. Hardware as a medium for integration.
6. Hardware as an embodiment of abstract concepts (functional and theoretical).
7. Hardware as an adversary.
8. Hardware as a prompt.
9. Hardware as a communication medium.

In observations of design activity, Harrison and Minneman (1996) found "that the processes of interaction with objects are an integral part of the communications, alter the dynamics in multi designer settings and form part of the pool of representations that are drawn on by designers" (role 1). This chapter describes roles 1–5. For more detailed descriptions and discussion of roles 6–9, see Brereton (1998).

Hardware starting points and kinaesthetic memory triggers

Hardware and prior experiences with hardware are the starting points from which students develop design proposals. As the previous section illustrated, students look for possibilities in existing hardware to meet design requirements. We might expect this result when hardware is readily available, as in the crane exercise. However, it is interesting to notice where students look for inspiration when there is no hardware at hand. In the scales exercise, students were asked to develop concepts for an internal mechanism for kitchen scales, as illustrated in Figure 4.2b, with only paper, pens, and a fully assembled kitchen scale which they were asked not to disassemble. The exercise revealed that in conceptual design, students draw on memory of experiences with hardware (internal representations) and are opportunistic in seeking out any kind of miscellaneous hardware (external concrete material representations) to think with. We can get some idea of students' internal representations by observing them in the conceptual design exercises. In recalling prior experiences with hardware, students mentioned winding clock springs and watching music boxes unravel. They recalled with varying success how moving coil galvanometers, pressure gages, wind up toys, and ball-point pen deploy-retract mechanisms work. It was notable that many groups, in saying that a design could be "like a biro" or "like a wind-up toy," did not make any explicit reference to the abstract function or geometry, but simply referred to behaviours of similar devices. Often these comments were accompanied by gestures or simple sketches in the form of recorded gestures that indicated that manipulating hardware led to bodily learning.

Table 4.1. The roles of hardware in mediating design negotiations and the associated learning outcomes

The roles of hardware in mediating design negotiations	Design learning outcomes
Hardware as a starting point	Hardware is tangible. It exists. It serves as a starting point, is easily noticed, remembered, seen and touched. It offers a basis for comparison. (It is a concrete external durable representation.)
Hardware as a kinaesthetic memory trigger	Episodes of kinaesthetic experiences with physical objects serve as memory devices (internal representations).
Hardware as a thinking prop	Hardware objects have all sorts of properties that afford different actions. Hardware that is easily accessible and has a useful property is adopted as a gestural aid to support thinking.
Hardware as a chameleon	Hardware is always in a context of use. What the hardware reveals depends on the context of use. A variety of informal experiments in different contexts reveal different facts.
Hardware as a medium for integration	Integrating components in their functional context reveals: practical limits of use; characteristics of operation; methods of connection; causal relations; and physical quantities. This empirical knowledge extends the student's hardware repertoire.
Hardware as an embodiment of abstract concepts (functional and theoretical)	Observing and testing hardware reveals through the hardware behaviour: fundamental concepts; physical embodiments of abstract concepts; and unanticipated design issues.
Hardware as an adversary	Challenging theoretical model predictions against hardware behaviour reveals discrepancies and provides clues to modelling errors. This reveals theoretical assumptions and causal relations.
Hardware as a prompt	Device behaviour prompts student questions and suggests experiments. Through repetitive interaction with hardware, students bring order, distilling out key operational parameters and their relationships.
Hardware as a communication medium	Hardware is integral to learning communications, affecting the course of enquiry, idea generation, discovery and the dynamics of group interaction. Hardware is used to command attention, to demonstrate, and to persuade.

The transcript below illustrates a typical conversation in which students drew upon their experience with hardware while designing scale mechanisms.

VIVIAN: If this is, em, you know, like one of those farmer toys where you pull the string and it rolls back, maybe it's something like that where it's maybe a spring-loaded coil or a spring-loaded, em, disk with a thing attached to it.

VIVIAN: Did you ever watch a music box unravel? Like, you know, these kinds of springs like this, so if you squish it, it causes some kind of rotation.

JUAN: Mmmm right.

VIVIAN: And if you have rotation in one orientation you can usually translate in one orientation you can usually translate it into another.

JUAN: Yeah, you're right. I guess I have watched those too.

VIVIAN: Yeah exactly right . . . right.

Figure 4.6 shows a design developed by one student who searched his bag to find a ball-point pen or biro, when he noticed the motion of the scale was similar to that of a biro. He proceeded to dissect the biro in order to learn how it worked. He sketched the mechanism. He then sketched a design for the scale that built heavily on the biro's deployment-retraction mechanism.

These observations provide evidence that one way in which we think about and remember ways of implementing abstract functions, such as linear to rotary motion, is through our experience of artefacts – that is, if we try to design a catch mechanism, one way to go about it is to seek inspiration from all sorts of things that we open and close: umbrellas, CD holders, doors, egg cartons, briefcases, laptop computers, VCRs, etc. Novice designers do not store a library of different kinds of abstract catch mechanisms where they remember the particular geometric configurations of each catch; whether or not experts do so is an open question. Rather, novices recall experiences of products that need catches to keep them open or shut. And they recall the catch in its particular context of use, remembering the feel of opening it.

Based on these observations we can draw some conclusions about the internal representations used by designers. At least, we can state how these memories are manifested as external representations on recall. Designers express what they remember about devices by:

1. Recalling the experience of using a device, noting the device behaviour, particularly referencing actions and movements.
2. Naming and drawing standard machine elements.
3. Gesturing and drawing geometric configurations of previously experienced devices.

Hardware thinking props

The second notable behaviour in the scale's conceptual design was the opportunistic seeking out and use of miscellaneous hardware to think with and gesture with (ready-made representations). In a barren design environment consisting of a classroom full of chairs, tables, sketch pad and pens, students sought out inspiration from: gesturing with pens; pulling and twisting a rubber band that happened to be lying on a table; and dissecting a ball-point

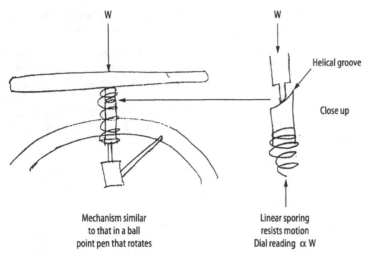

W   W

Helical groove

Close up

Mechanism similar
to that in a ball
point pen that rotates

Linear sporing
resists motion
Dial reading α W

Above: A sketch taken from a student's sketch pad showing a design based on a biro (ballpoint pen) deploy-retract mechanism.
Below: A design conversation in which students build on the design sketched above.

Raul: [Looks in bag for a biro (ballpoint pen).] I've had the experience of taking apart a biro. I reckon it could be like a biro.
Mark: [Laughs] You reckon it could be like a biro?
Raul: It could be when you think about it.
          [Examines biro and sketches for a while.]
Mark: What's there?
Raul: That's a close-up of that area there (see sketch above). It's like a pen – you know, how one of these pens as you're pushing it down that's got those tags in it and they make this go around, like when you put that in it pushes round and so that rotates it should do that and I was thinking like it could be like that with these grooves and, if instead of having gaps, you have like one spiral groove there, you could press this down and this is spring-loaded at the bottom for resistance and so, however much you push this down, this rotates an amount (inaudible).
Mark: That' s really cool Raul.
Liam: Novel, novel.
Mark: I dig that, that's good.
Mark: So is it meant to be keeping on going round so that.
Raul: Yeah, enough so that.
Mark: I suppose you can only go around once. Well, no, you could have it going around more if you had this pushing down from the outside.
Raul: If you had it winding in a taper or something.
Raul: Like if your helical groove thing was like and you had your tongue thing sitting out here on the groove like a screw thread that screws it around.

Figure 4.6 Hardware as a starting point and memory device – designers build on experience with existing hardware devices.

pen dug out from a student's rucksack. Of the incidental hardware in the room, students adopted hardware tools that were easily accessible and that had affordances or convenient properties. Pens were long and slender like linkage links and rubber bands were stretchy like springs. This is not to say that the properties were optimal or entirely suitable, or that it would be possible to specify a priori what would be suitable. The hardware was simply conveniently available and had some attribute that made it helpful for students to gesture and think with. This opportunistic behaviour has been observed in

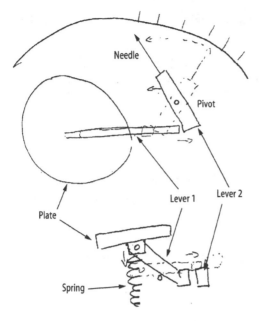

Left: Student's concept sketch of a kitchen scale designed by gesturing with pens.
Below: Transcript of a student describing his group's design to the class. While describing the design, the student gestures with pens and points at the sketch. Typed labels and arrows are added for clarity by the author.

"I was working on the kitchen sacle design.

Ther's a spring there and there's a plate pushing down on top, OK? And that's pushing down this lever which, as that is pushed down on [the top] end, this [lower end] is moved out.

And that movement in that direction is pushing the base of another lever here which is pivoted over here. So, as that moves out, that moves the needle around. So it's a really simple design."

Figure 4.7 Hardware as thinking prop – students appropriate convenient hardware to think and experiment with.

longer projects. It lends support to the idea that an accurate model is less important than a quick model that helps to explore the space.

Figure 4.7 shows a design that was developed by a group who gestured with pens in their hands to develop a linkage mechanism. The pens formed the two levers of the linkage mechanism.

Hardware as a chameleon

Hardware relies on its context of use for its functional meaning. A paper-clip may be seen as:

- a device to hold pieces of paper together;
- a thin piece of wire for picking a lock;
- a thin piece of wire for pressing a recessed button to restart a computer;
- an electrical conductor.

There is a sense in which hardware function is indexical,[1] in that it relies on its context of use for its particular functional meaning. It is analogous to language in the way that its meaning changes with its context of use.[2] Brown et al. (1989) argued that machines indexicalize abstract representations, pointing out that "in an intriguing way you need the machine to understand the manual as much as you need the manual to understand the machine". In fixing my own old car I have found it helped to have several manuals in addition to the car. The car helps me to interpret the manuals and each manual helps to interpret the other manuals.

Abstract representations, such as engineering drawings or theoretical concepts, on the one hand, have conventional meanings which hold on any occasion of their use, but, on the other hand, they draw their meaning from the physical world. They are derived from and understood by the variety of specific physical scenarios to which they apply. In learning to use abstract representations it is thus necessary to understand and discover them in the context of specific physical scenarios.

Hardware as a medium of integration

Integration, the bringing together of different components to synthesize a new whole, distinguishes design from analysis. Integration typically reveals all manner of issues that cannot be anticipated. It is a great source of learning.

When designers integrate components in a design, they learn both about the individual components and the interfacing issues that ensure that one component works properly with the next. Integrating components in their functional context reveals their:

- practical limits of use;
- characteristics of operation;
- methods of connection;
- causal relations;
- independent variables; and
- physical quantities.

This empirical knowledge extends the designer's hardware repertoire[3] (Schön, 1994).

Developing the hardware repertoire through integration

As designers explore ways to meet design requirements, they expand their hardware repertoire. By a "hardware repertoire" I mean a repertoire of components, materials, synthesis skills and associated design contexts with which designers have experience. When designers "add" a component to their design repertoire, they develop the knowledge of its characteristics and the limits of its behaviour in the context of use. They develop this knowledge through attempting to integrate the component into a design, with a design goal in mind. The knowledge that results from this attempt then forms part of the repertoire on which they draw in future design projects.

I will illustrate the notion of learning through integrating components with an example drawn from a two-week project to design an efficient model all-terrain vehicle. This vehicle must cross a stretch of gravel and climb a carpeted ramp using the least amount of energy possible. The design project necessitated that students negotiate between abstract and material representations, as must be done in professional engineering design practice. They had to select motors from data sheets and had to present quantitative experimental results in a design document, as well as build a hardware device. Table 4.2. summarizes a variety of student discoveries that occurred through hardware integration in the all-terrain vehicle project. Many of the discoveries are very specific. However, this specific empirical knowledge supports the understanding of concepts. Developing a useful understanding of a formal concept relies on discovering it in several different real contexts. The variety of contexts bounds the concept, establishing its characteristics, limits and conditions of use. The kinds of learning detailed in Table 4.2 – practical limits, methods of connection, causal relations, etc. – fall into a similar set of categories as the errors identified by Miller (1995) in student work.

The conversation presented in Figure 4.8 illustrates the students' level of understanding at the early stages of the project. In the course of the project these students made all the discoveries listed in Table 4.2. In Figure 4.8 the students are discussing how the vehicle could detect when it is on the ramp because they "might need to shift gears to go up the ramp." They have heard of terms such as "stall" and "no-load speed" in their lectures, but these have

Table 4.2. Examples of knowledge acquired in the all-terrain vehicle project

Learning outcome	Examples of specific discoveries made through hardware component integration
Practical limits of use	Motor stops at stall torque limit, current consumption is high, wheels won't turn, motor gets hot. Motor coils physically burn out if run above rated voltage for long.
Characteristics of operation (properties of materials and components)	Motor gets hot at stall. Torque speed curve is linear for permanent DC magnet motor. Voltage supplied by 3V alkaline batteries gradually drops from about 3.2V to about 3V, but batteries lose capacity to supply current.
Methods of connection	Motor must be bolted to chassis to apply reaction torque. Lego spline shaft must be attached to motor shaft concentrically or needs flexible coupling to attach motor shaft to load.
Causal relations	Reversing current reverses permanent DC magnet motor direction. As wheel size increases, motor torque needed increases or gearbox must be adjusted. Lighter vehicle puts lower load on motor, lower forces in gearbox, less friction in gearbox. Lighter vehicle requires less power to move it.
Independent reference variables	Work done moving vehicle load up incline is constant. Load on motor from vehicle (through gearbox) determines motor operating speed and motor torque supplied. Must change vehicle weight or change gearbox to change load at motor. Battery voltage is roughly constant. Current drawn depends on impedance.
Physical quantities	50% losses in LegoTM gearbox due to friction. Particular 3V hobby motor draws about 100mA, supplies torque of 1in oz. Two AA alkaline batteries running 3V motor drawing 100mA last approximately 10 minutes.

Motor 1 Torque-Speed Curve

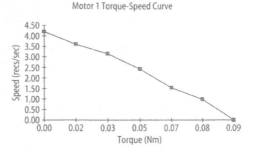

1 Carol: We might need to shift gears to go up the ramp. It could need more torque.
            Could it sense its wheels don't work? You know when the torque is . . .
2 Alice: You mean at stall. [The team looks impressed.]
3 Sean: You listen in class.
4 Alice: At stall it sucks up a whole lot of current. It could blow a fuse.

Figure 4.8 Building on the hardware repertoire. Conversation of three students designing a model motorized all-terrain vehicle as they work to make sense of abstract representations in the context of prototype hardware.

been used in the context of equations and abstract representations; the lecturer explained the concept of stall using a torque-speed graph and the idea that current consumption increased with torque by using a torque-current graph. However, the students have only just begun to try to apply these concepts in the context of hardware design. If one has never tinkered with or pondered about motors, it takes some thought to grasp the idea of the motor reaching a limit where the motor speed reduces to zero and the motor cannot supply enough torque to drive the load. It takes some effort to link the abstract representations on the board to the performance characteristics of hardware and to the problem at hand, and to internalize the meanings of the abstractions – i.e., to relate them to physical experiences. As the conversation in Figure 4.8 indicates, two of the three students initially had difficulty connecting the concept of stall to the problem at hand.

Notice that in phrase 4 of the conversation in Figure 4.8, Alice uses her knowledge of the characteristics and limits of motors to make a rudimentary design proposal. Knowing that the motor stalls at a certain torque and that the motor draws more current when stalled, she proposes that the motor could blow a fuse in order to activate some means of climbing the ramp. Through the process of identifying the characteristics of components, making design proposals and synthesizing them, students gradually build up their hardware repertoire.

After some conceptual design discussions and consultation with the professor, the group began prototyping. As they developed the prototype they gradually identified the operational characteristics and limits of the motor, using transient representations such as speech and gesture to refer to them and then reconfiguring the hardware to account for what they had learned. They decided to begin with a fixed transmission for simplicity. Working together, Carol and Sean hooked up a motor to their vehicle drive train. Neither had connected up a motor before. They positioned the motor on the chassis, meshing the gear attached to the motor shaft with the larger gear on the vehicle's front axle. As they connected power to the motor, which was resting on, but not secured to the chassis, they were visibly surprised as the motor leapt off the chassis. They had not anticipated that the motor must be secured in order to react to the motor torque through the chassis. Having secured the motor to the chassis and connected it to the wheel axles through a gear train, they switched on power again. This time they noticed that the wheels spun when the vehicle was held in the air, but that the vehicle would not move when placed on the ground. They identified correctly that they needed more torque at the wheels and added another gear stage to the transmission. They pointed out the relative speed of the wheels in the air compared to their speed along hard and carpeted floors. They commented when it stalled in various conditions. They hooked it up to a multimeter to find out how much current it drew from the batteries. When the leads came off and they hooked them back up, they suddenly noticed and remarked that the motor was going in the other direction. They deduced that they had reversed the current direction and then reversed the leads to check that the motor turned the opposite way again. They noted that for this kind of motor, the direction of the current mattered.

Through synthesis and testing the students made several empirical discoveries relating to torque, current, speed and stall, listed in Table 4.2. They got a "feel" for motors, developing an understanding of their characteristics and limits of use. Even though Sean and Carol did not make a connection to the lecture material in the first instance, they gradually did so throughout the course of the exercise. They came to understand that the applied load was the reference variable that controlled motor speed; that too much load leads to stall, etc. They learned about abstract concepts through integrating physical components because successful physical integration demanded that they understood the operational characteristics of components under specific interfacing conditions. Furthermore, the knowledge that they gained is based in a design context – the next time the students encounter stall, they are likely to think of their all-terrain vehicle stalling under different conditions. Having had such experiences and developed empirical knowledge of the motor's performance characteristics and physical limits, the students are developing, one can presume, the kind of experiential knowledge that is associated with experts who have many previous design projects to draw on – that is, they are expanding their hardware repertoire through integrating components.

In summary, students develop a large amount of empirical knowledge in designing that supports the understanding of concepts. General formal concepts exist precisely because there are a large number of real contexts in which they apply. However, developing a useful understanding of a formal concept relies on discovering it in several different real contexts. The variety of contexts bounds the concept establishing its characteristics, limits and conditions of use.

General remarks

Different representational frameworks are based on different conventions and underlying assumptions. By challenging one representation against another, a designer can uncover gaps in thought. The underlying assumptions of the different representational frameworks are often brought to the fore. For the engineering designer it seems to be particularly fruitful to challenge abstract against material representations. This process brings to light new information about hardware characteristics, the design requirements or the working explanatory model. In other disciplines it may make more sense to challenge one kind of abstract representation against another.

It is interesting to note that many successful theoreticians, notably Feynman and Tesla, mention in their autobiographies that they were childhood tinkerers. Other research work (Brereton 1998) has shown that successful students develop their understanding of unfamiliar fundamental physical concepts through habitually using them to try to explain the behaviour of physical devices. Once these fundamentals are well understood, these students can then extend their understanding to much more complex and abstract technical domains, with only the occasional need to re-represent and test ideas in the physical domain. They can also test ideas developed in one abstract representational framework through the use of a second abstract representational framework. However, one cannot develop a theoretical understanding of the physical world without these early experiences of constantly challenging abstract models against the physical world. This process ensures that assumptions, concepts and relations are encoded correctly by testing, correcting and verifying them in a variety of scenarios.

Conclusion

This chapter has presented a qualitative analysis of learning in design discussions, paying particular attention to how learning in design arises from negotiating between abstract and material representations. Abstract representations are derived from and understood by the variety of specific physical scenarios to which they apply. In learning to use abstract representations, it is thus necessary to understand and discover them in the context of specific physical scenarios. The process of design always benefits from a variety of representations. Shifting between representations in order to understand how to close the gap between representations advances understanding of the design.

Acknowledgements

Much of this work was funded by the National Science Foundation Synthesis Coalition. Data gathering was conducted during my PhD studies at Stanford University Design Division. I would like to acknowledge the contributions of Professor Larry Leifer and to thank researchers at Stanford Centre for Design Research, The Institute for Research on Learning and XEROX PARC Design Studies Group, all of whom discussed videotapes with me. I am very grateful to Gabi Goldschmidt, William Porter, Yvonne Rogers, and Pamela Siska for helpful suggestions on the manuscript. The Information Environments Program at the University of Queensland has provided a fertile environment to continue exploring this work.

Notes

1. Expressions that rely on their situation for significance are commonly called indexical, after the "indexes" of Charles Pierce (1933) (from Suchman, *Plans and Situated Actions*, p. 58).

2. Suchman (1987), in *Plans and Situated Actions*: (The Problem of Human-Machine Communication), p. 58, writes: "The efficiency of language is due to the fact that, on the one hand, expressions have assigned to them conventional meanings which hold on any occasion of their use. The significance of a linguistic expression on some actual occasion, on the other hand, lies in its relationship to circumstances that are presupposed or indicated by, but not actually captured in the expression itself. Language takes its significance from the embedding world, in other words, even while it transforms the world into something that can be thought of and talked about." Heritage (1984, p. 143) offers as an example the indexical expression "that's a nice one," pointing out that the significance of the descriptor "nice" has a different meaning if it refers to a photograph or to a head of lettuce.
3. Schön 1994 has used the term repertoire in describing architectural design practice. Here I relate the idea of a repertoire to mechanical engineering learning. I pay particular attention to how the repertoire grows as a student implements a new piece of hardware in a context and in so doing learns about its conditions and limits of use.

References

Brereton, MF 1998. The role of hardware in learning engineering fundamentals. PhD dissertation, Stanford University.

Brereton, MF, DM Cannon, A Mabogunje, and L Leifer 1996. Collaboration in engineering design teams: How social interaction shapes the product. In Analyzing design activity, edited by N Cross, H Christiaans, and K Dorst. Chichester, UK: John Wiley & Sons, pp 319–341.

Brereton, MF and LJ Leifer 1997. The role of hardware, context and noise in learning engineering fundamentals. In Proceedings of the International Conference on Engineering Design Volume 2. Edited by A Riitahuhta. Tampere, Finland, pp. 591–596.

Brown, JS, A Collins, and P Duguid 1989. Situated cognition and the culture of learning. Educational Researcher 18(1):32–42.

Chaiklin, S and J Lave 1993. Understanding practice: Perspectives on activity and context. Cambridge: Cambridge University Press.

Dourish, P 2001. Where the action is: The foundations of embodied interaction. Cambridge, MA: MIT Press.

Goodwin, C 1995. Seeing in depth. Social Studies of Science 25:237–74.

Harrison, S and S Minneman 1996. A bike in hand: A study of 3D objects in design. In Analyzing design activity, edited by N Cross, H Christiaans, and K Dorst. Chichester: John Wiley & Sons, pp 417–436.

Heritage, J 1984. Garfinkel and ethnomethodology. Cambridge, NY: Polity Press.

Hutchins, E 1995. Cognition in the wild. Cambridge, MA: MIT Press.

Jordan, B and A Henderson 1995. Interaction analysis: Foundations and practice. The Journal of the Learning Sciences 4(1):39–103.

Macann, C 1993. Four phenomenological philosophers. London: Routledge.

Miller, CM 1995. So can you build one? Learning through designing: Connecting theory with hardware in engineering education. PhD thesis, MIT.

Moll, LC 1990. Vygotsky and education. Cambridge: Cambridge University Press.

Pierce, C 1933. Collected papers, Vol 2. Edited by C Hartshorne and P Weiss. Cambridge, MA: Harvard University Press.

Schön, DA 1983. The reflective practitioner: How professionals think in action. New York: Basic Books.

—— 1994. Designing as a reflective conversation with the materials of a design situation. In Research in Engineering Design. Volume 3. New York: Springer Verlag, pp 131–147.

—— 1990. The design process. In Varieties of thinking, edited by VA Howard. London: Routledge, pp 100–141.

Schutz, A 1932. The phenomenology of the social world. Evanston: Northwestern University Press.

Suchman, L 1987. Plans and situated actions: The problem of human machine communication. Cambridge: Cambridge University Press.

5

Design Representations in Critical Situations of Product Development

Petra Badke-Schaub and Eckart Frankenberger

Introduction and Objectives

Industrial product development means complex problem solving by mostly interdisciplinary and sometimes even intercultural teams (see Pugh 1990; Ehrlenspiel 1995; Pahl and Beitz 1997). Thereby, the team members interlink various pieces of design information in a creative process to achieve a final product (Cross and Cross 1996). This exchange of information takes place in different types of design situations and includes different activities such as searching, analyzing, selecting, evaluating, emphasizing or the dropping of information. Of course, we can also distinguish among different types of content of information, such as test data, customer requirements, and performance numbers (see Hales 1987; Blessing 1994; Frankenberger 1997). Moreover, design information is transferred by various representations, including verbal, written, sketches, drawings, and electronic data. The chosen representation of design information has to suit many requirements of very different design situations in order to provide the necessary availability of information. This availability of information is revealed to be a central factor in the success of design work (Badke-Schaub and Frankenberger 1999). Consequently, the use of internal (in the designer's mind) and external representations, as well as switching between both types of representation, are the essential mechanisms of thinking and acting in engineering design (Dörner 1998; Hacker, Sachse and Schroda 1998).

The procedure of searching, furnishing, and evaluating information depends to a large degree on the situation itself, the type of information, and individual prerequisites such as knowledge and experience. In order to support and improve design processes, we have to understand the central mechanisms that determine this complex process of information transfer. An approach limited to the needed content of information cannot be sufficient and we have to consider the representation of the design information, too. In this chapter we introduce selected results that are related to the external representations of design information. These results were derived from an empirical study of the engineering design processes in industry (Badke-Schaub and Frankenberger 1999). We concentrate on the verbal representations of design information transfer in critical situations of the design process.

Methods

Investigating the requirements of information representation in design necessitates a detailed analysis of the designers' activities. But there is more to compile than information on the activities that occur during the design process. The aim of a differentiated understanding of information representation in design practice implies dealing with a large number of dependent and independent variables from different fields. Thus we based our investigation on a general model of four central influences on the design process in practice: "individual prerequisites," "prerequisites of the group," "external conditions," and the "task" (see Figure 5.1, with examples).

The complexity of such an approach makes it impossible to investigate so-called "comparable groups" in several companies. On the contrary, the investigation has to focus on a very detailed observation of "single cases"

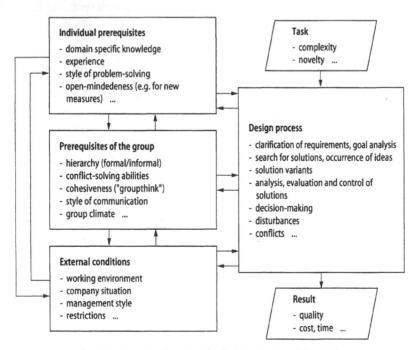

Figure 5.1 Factors influencing the design process and the result.

over an extended period of time. Therefore, we chose a single-case approach with two investigations in industry.

The first investigation took place in a company producing agricultural machinery. Over the course of four weeks, the design process of a group of four designers redesigning a fruit press was observed and documented (see Figure 5.2). The second investigation was conducted in a company of the capital goods industry. In this company we observed three projects of a design team developing and redesigning several components of a particleboard production plant (see Figure 5.3) for eight weeks.

Figure 5.2 Pneumatic fruit press.

Figure 5.3 Particleboard production plant.

The large number of influencing factors requires the use of a variety of investigation methods. The following sections describe the methods used to compile information on the "external conditions," the "design process," the "individual prerequisites," and the "prerequisites of the group." For the analysis of design representations, the data concerning the designers' activities are of specific interest. The compilation of factors from the individual and the group will be important in a second step for additional analysis of the underlying mechanisms (see Figure 5.14).

External conditions and the design process

Several different aspects regarding conditions were recorded, such as branch, the economic situation of the company, its culture and organization, the flow of information and communication within the organization, and, last but not least, the direct working environment.

To compile the dynamic course of the design process, a detailed analysis of the design work at short time intervals is required. The duration of intervals was determined by the characteristics (categories) of the communication in order to describe the process. To this end a standardized approach for investigating cooperative design in industry was developed by combining direct and indirect methods.

The primary direct method was continuous non-participant observation, involving two observers – a mechanical engineer and a psychologist – sitting in the same room as the designers. The mechanical engineer observed the activities of the designers in terms of, for example, working-steps in accordance with those used in the systematic design approaches seen in Pahl and Beitz (1997), and the development of technical solutions in terms of subfunctions/components, ideas and solution variants. The psychologist focused on cognitive and social aspects such as decision-making and group interactions. A standardized laptop-based "online" protocol was used to document the observations in real time. This protocol provided a first description of the design work as a problem-solving process. Video recordings of all teamwork and the relevant phases of individual design work were used to review and obtain a detailed account of specific interesting phases of the individual and the group design process (cf. Frankenberger and Auer 1996).

The final protocol of the observed design process consisted of a word-by-word transcription of important dialogues (see Table 5.4) and a description of the entire process with an average duration of 30 seconds per protocol line. These protocols formed the material for a qualitative and quantitative analysis of the process, using special software that allows easy analysis by presenting graphic representations of each process characteristic against time. These graphs represent the development of the solution by showing the moves between the sub-problems and solution variants.

In addition to these direct methods, indirect methods, such as diary sheets, were used. Diary sheets were used as a basis for short semi-structured interviews each evening in which the designers were asked about their successes and failures of the day, how they solved problems, and when and why they contacted their colleagues. The diary sheets were designed to be filled out by the designers with minimum effort, in order to avoid loss of motivation.

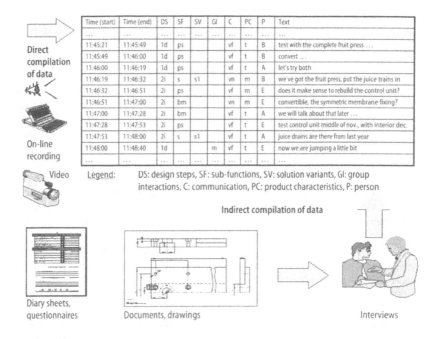

	Time (start)	Time (end)	DS	SF	SV	GI	C	PC	P	Text
	...	...	...	...	...	...	...	...	...	...
	11:45:21	11:45:49	1d	ps			vf	t	B	test with the complete fruit press ...
	11:45:49	11:46:00	1d	ps			vf	t	B	convert ...
	11:46:00	11:46:19	1d	ps			vf	t	A	let's try both
	11:46:19	11:46:32	2i	s	s1		vn	m	B	we've got the fruit press, put the juice trains in
	11:46:32	11:46:51	2i	ps			vf	m	E	does it make sense to rebuild the control unit?
	11:46:51	11:47:00	2i	bm			vn	m	E	convertible, the symmetric membrane fixing?
	11:47:00	11:47:28	2i	bm			vf	t	A	we will talk about that later ...
	11:47:28	11:47:53	2i	ps			vf	t	E	test control unit middle of nov., with interior dec.
	11:47:53	11:48:00	2i	s	s1		vf	t	A	juice drains are there from last year
	11:48:00	11:48:40	1d			m	vf	t	E	now we are jumping a little bit
	...	...	...	...	...	...	...	...	...	...

Direct compilation of data

On-line recording

Video Legend: DS: design steps, SF: sub-functions, SV: solution variants, GI: group interactions, C: communication, PC: product characteristics, P: person

Indirect compilation of data

Diary sheets, questionnaires

Documents, drawings

Interviews

Figure 5.4 Compiling the design process using direct and indirect investigation methods.

Additionally, the documents produced by the designers (drawings) were also collected, and they were asked about their processes and their results. These interviews, based on the diary sheets and the documents, provided important information about the design process and helped us to understand the development of the solutions and the technical decisions. Figure 5.4 depicts the procedure of compiling data on the design process and presents an excerpt of a revised on-line-protocol.

Individual prerequisites

Individual behaviour (e.g., communication generated by a person) is influenced by several factors. A reduction of the complex cognitive, emotional, and behavioural processes to one or two "important" characteristics seems almost impossible. People usually behave according to the situation at hand: few paradigms can be considered universally valid for all situations or all individuals' behaviours. For example, a person confronted with a novel, complex problem will take longer to analyze it if there is enough time, if the problem is important or if there seems to be a good chance of solving the problem, than he or she would in a situation in which there is no time or the problem is less important. Therefore, different methods were chosen in order to assess the individual prerequisites (see Table 5.1).

Biographical data and personal opinions about the working environment were mainly collected by means of semi-structured interviews. Assuming that design processes are fairly typical examples of complex, realistic problem-solving processes, it is important to look at the engineers' strategies in complex and novel situations. The designers' ability to deal with complex problems was assessed by analyzing the thinking and action-regulation behaviour of each designer while solving computer-simulated problems (cf. Dörner and Wearing 1995). Each designer was asked to solve two problems that were novel, complex and dynamic. These simulations were selected because they required different manners and strategies of action regulation. Contrary to design tasks, computer-simulated problems can be solved without any specific textual experience. The behaviour of the subject is not measured as a single numerical variable (e.g., the "quality" of problem solving); instead, the focus is on the action-regulation styles of the individual (i.e., the planning process of the subject), consisting of sequences of different variables such as the evaluation of questions, decisions, etc. In using these standardized computer-simulated problems, individual heuristics and strategies can be investigated (Badke-Schaub and Tisdale 1995). Other studies have shown that the strategic behaviour of designers in these simulated problems is similar to behaviour in design work. These similarities can be interpreted as individual action styles (Eisentraut 1997).

The assessment of the heuristic and social competence of the designers was based on their design process (captured in the final protocols and the diary sheets) and on a self-assessment questionnaire developed by Stäudel (1987). Several studies on heuristic competence indicate that a positive self-assessment of problem-solving abilities supports successful problem solving in complex situations (cf. Stäudel 1987). The social competence of the designers was assessed using the observations of group activities, both during the design work and during the work with the computer-simulated problems.

Table 5.1. Variables and methods for compiling individual prerequisites

Field of data	Variables	Methods
Biographical data	Age Professional education, career Qualifications and experience	Semi-structured interview Questionnaire
Work environment	Motivation; job satisfaction Evaluation of the organization Evaluation of the actual project Relationship to colleagues and to superiors	Semi-structured interview Questionnaire
Ability to deal with complex problems	Analysis and information-gathering Action planning Dealing with time pressure Dealing with stress	Computer-simulated micro-worlds Fire (individual) Machine (individual) Manutex (group)
Competence	Heuristic competence Social competence	Questionnaire (Stäudel, 1987) Observing and analyzing the inter-actions of the group
Abilities concerning the design process	Clarification of the task Search for conceptual solutions Selection and control	Diary sheets/marks-on-paper On-line protocol of the design process (Video and tapes)

Group prerequisites

The aim of collecting group prerequisites was to investigate the structure and the organization of the group during the problem-solving process and to investigate how the group approached the problem in terms of behavioural patterns that may be responsible for producing the observed design representations. It was decided to focus on group interaction processes and to describe these in terms of individual and group behaviour patterns. Consequently, we chose phases of group interactions during the design processes and described them in terms of individual and group behaviour patterns. Another important diagnostic situation was a third computer-simulated problem that was given to the designers as a group.

Whereas the problem-solving activities demand a high degree of goal-analysis and emphasizing of priorities, the group situation necessitates that each individual expresses his or her own ideas and strategies of proceeding. Getting his or her own suggestions accepted is linked to the different characteristics of the individual – mainly, the concept of social competence, which includes several abilities of acting in groups (e.g., the ability to cooperate and the ability to communicate).

The results of the computer simulation were compared with the results of specific periods in the observed design process. The same encoding system was used in both cases, based on the phases of action regulation developed by Dörner (1998). Additionally, socio-emotional behaviour and organizational aspects were categorized.

A summary of the data of the different elements of the initial model – the domains of influencing factors – is given in Table 5.2.

Selected Results

The analysis of the design processes can proceed in different directions. First, attention can be directed to limitations at the organizational level. Second, the different motivational and cognitive processes of the individual, which play an important role during the whole process, can be addressed. Third, forces that have effects on the modes of communication in the group can be investigated.

Table 5.2. Methods for compiling data on the elements of the initial model

Methods	Domains of influencing factors			
	Design process and result	External conditions	Prerequisites of the individual	Prerequisites of the group
Interviews	●	●	●	●
On-line-protocols	●	●	●	●
Diary sheets	●	●	●	●
Marks-on-paper	●		●	
Questionnaires			●	●
Computer-simulated problems			●	●

Following our investigations, we accumulated extensive data on design work that allowed us to analyze several result- and process-related questions. Concerning the role of information transfer in the design process, we have three questions. First, what are the situations in which information transfer has the greatest impact on further processes? Second, which design representation is the most common form of information transfer in these situations? Third, how do specific kinds of information transfer influence further design processes and the final design result, in terms of success and failure? It is in terms of these questions that we present our analysis in the following sections.

When does information transfer have the greatest impact?

The analysis of design work reveals that not every moment is of the same importance for subsequent design activities and for results obtained later. For example, at certain moments we may observe designers sitting at their CAD station adding holes and screws during the embodiment design phase of a component; at other moments they are engaged in deciding on concepts or solution principles. Obviously, we can distinguish between "routine work" and important "critical situations" that determine "choice points" in the process and thus in the whole project. These critical situations underlie the remaining design process in a positive or negative way. Consequently, critical situations are of specific importance to the success of the design process.

Critical situations are therefore of special interest in isolating the main influences on the design process. In order to extract these influences and explain the effect of a critical situation, we developed a submodel of the interdependencies between the influencing factors and the process characteristics for each critical situation (see Figure 5.5). Evidence for each identified relation was gathered separately. Special interviews with the designers, combined with video-feedback of selected critical situations, helped us to revise the submodels.

The sum of the different interrelations in the individual submodels led to a model of relations between influencing factors and process characteristics in all critical situations of the design process. Altogether, 265 critical situations

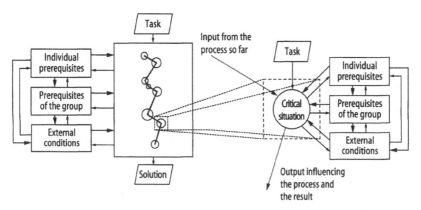

Figure 5.5 Influences on the design process as influences in "critical situations."

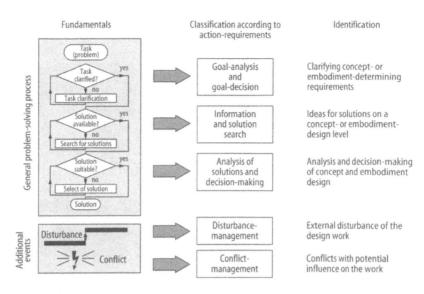

Figure 5.6 Division of "critical situations" according to the general problem-solving process and additional events (social context).

were identified in the four analyzed projects of the two investigations. These explained the course of work through more than 2200 single interrelations between factors, process characteristics, and the result. A reduction to 34 different influencing factors illustrates the suitability of the model (see Frankenberger and Badke-Schaub 2000).

Derived from the steps of general problem solving, we can contrast different types of critical situations that regard their aim in the problem-solving process, such as goal-analysis, goal-decision, solution-search, solution-analysis, and solution-decision. Moreover, we can observe situations that are important in their social context, such as "conflicts" and "disturbances." These situations require "conflict-management" and "disturbance-management."[1] (see Figure 5.6)

Next, we focus on the following question: what kind of information transfer is most commonly used and what kind is mostly successful?

What kind of information transfer is used in critical situations?

In our investigations, the designers were working individually about 70% of the entire working time. However, nearly 90% of critical situations occurred during instances of collaboration. As illustrated in Figure 5.7, communication between colleagues is extremely important for the exchange of decisive design information.

In a different study, interviews in ten R. & D. departments of major German companies underline the importance of colleagues as the most frequently mentioned source of information transfer in everyday design work and verbal communication was described as the most important mode of design representation (Badke-Schaub and Frankenberger 1998). Designers reported that

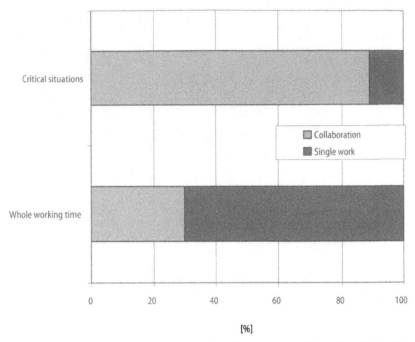

Figure 5.7 Amount of direct communication with colleagues in general (working time) and in critical situations.

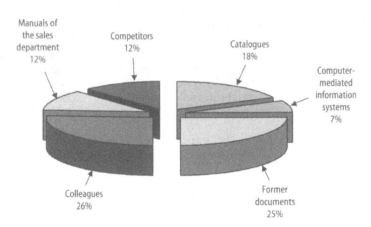

Figure 5.8 Percentage of use of different information systems by designers in their daily work.

26% of their own information-seeking processes are driven by asking colleagues, and only in 7% of cases do they search for information in computer-mediated information systems (see Figure 5.8; Badke-Schaub, Stempfle and Wallmeier 2001).

The heavy emphasis on verbal communication is surprising if we keep in mind that drawing and sketching are said to be the designers' "language." In an interview study Görner (1994) analyzed commentary by 74 experienced designers. The question they answered was whether they use – in order to develop a solution principle – "mainly thinking about it or mainly sketching." Of the designers, 69.3% were mainly sketching, 3.8% were "mainly thinking" and 26.9% reported a balanced combination of thinking and sketching. This result, however, is only valid for individual work, whereas engineering design projects are usually carried out by teams of several designers and thus a high degree of collaboration is required; therefore, verbal communication is a major part of designing in everyday work.

What is the reason that verbal communication is the main venue for "design representation" for the purpose of information transfer in decisive "critical situations"?

How are verbal design representations initiated?

We started our analysis of design representations by asking the question: what triggers the exchange of design information? To better answer this question, we differentiate between "active" information requisition and "passive" reception of important information.

Figure 5.9 depicts how often focused questions to colleagues on the one hand and individual "independent" information searches in documents on the other, were observed in critical situations. Results show that individual searches in documents were much less successful in terms of their impact on the result than focused questions. It is not surprising that, contrary to the active search for information, the passive reception of information always accompanies the good availability of this information. Passive reception of information means that the information transmitted was not asked for. This information transfer can happen unintentionally as part of a (often informal) conversation between colleagues, or it can be initiated by a colleague who is interested in informing a co-worker who is likely to need this information later in the design process.

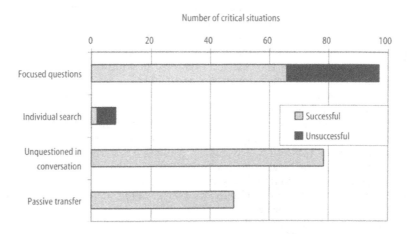

Figure 5.9 Successful and unsuccessful ways of information transfer in critical situations.

In connection with the high percentage of collaboration in critical situations, it is interesting that, in these situations, individual search for information in documents occurred very rarely and was not successful for the most part. But why is this the case? People do not like to search for information in catalogues, in computer-mediated systems, or in information lists because they assume that the search will take too long and because their previous experience has been that such a search often had not elicited relevant information. A further strong argument for preferring consultation with colleagues is that it provides an additional check on the context and the content of the required design information.

Another interesting aspect is that not only formal and organized information transfer is helpful; informal conversation can also be very useful: we observed that nearly 30% of important information is delivered "unsolicited" through informal conversation, in addition to the information transfer that follows a specific question. The great importance of verbal communication in determining information availability motivates us to take a closer look at verbal design representations in different types of critical situations.

What are important aspects of information transfer in critical situations?

Observing the verbal exchange of design information in design teams, we realized how diverse the ways to "shared understanding" may be (Klimoski and Mohammed 1994). Therefore, the important questions for us are reiterated: what kind of communication is suitable in a specific situation and, consequently, what kinds of communication support a successful result?

Without any doubt, the basic elements of communication, such as listening, positive feedback etc., are relevant to any form of interaction, including in a design team. We must keep in mind, though, that addressing negative aspects of a design proposal or solution, presented by a colleague, risks hurting the feelings of the colleague(s) and hence threatening the group climate. Because of this conflict, group-training often concentrates on supporting group climate in order to increase the effectiveness of group performance.

Our main goal was to determine the central elements of information transfer. Therefore, we created a category system that divides information into four categories: questions (searching for information), explanations (furnishing with information), evaluations (positive and negative statements); and procedural instructions (process information). In addition, there are three "neutral" categories: silence, repetition, and "other."

We encoded 11 positive and 11 negative critical situations of the types "solution-search," "goal-analysis and goal-decision," and "solution-analysis and solution-decision," comprising 2318 entities (sentences = communication units). Table 5.4 is a sample of the encoded transliteration of an informal conversation between two colleagues, B and C. It is derived from a critical situation of the type "search for a solution" during which the two designers were searching for a roller bearing for the spreader head of a particleboard production plant. The first column displays the time of the utterance, the second column shows the speaker, the third column transcribes the utterance, and the last column shows the assigned category applied to describe the information process (for abbreviations see Table 5.3).

Table 5.3. Categories of information transfer

Categories	Differentiation	Cat
Searching for information	All kinds of questions	aa
	Demands for attention ("Did you consider . . .")	ab
Furnishing with information	Ascertainment, conclusions	if
	New information	in
	More precise statements, specifications	ip
Evaluating: positive statements	Unspecific/ general positive statements, affirmations	pz
	Positive statements with regard to design aspects	pk
Evaluating: negative statements	Unspecific, general disaffirmation, rejections	na
	Negative statements with regard to design aspects	nk
Process information	Concerning further proceeding with regard to design aspects	vk
	Concerning further proceeding with regard to the organizational context	vo
Helplessness, silence	Mostly occurring after longer discussions	s
Repetitions	Iterations of just given information	w
Other	Other utterances	r

The dialogue in Table 5.4 captures a discussion of different ideas for designing a roller bearing. Both designers are experienced and have been working together for many years in a climate of high confidence. Of course, this situation is a particular case of information transfer between two specific designers and each individual case is peculiar to the persons, situation, group, and problem in question. Nevertheless, the exchange above is a typical example of information transfer processes during critical situations and it serves to reveal central influences that design representations have on the success or failure of the process to yield positive results. Analyzing different types of critical situations with respect to information transfer, we were able to detect several types of representation or patterns of communication that are positively correlated with a more or less successful result (Badke-Schaub 1998). Below we concentrate our analysis especially on critical situations of the type "solution search" because these are the "magic moments" in which new solutions are revealed and connected with "creative thinking" in design.

First we analyze the percentage of new information in the critical situations of the "solution search" type as compared to other types of critical situations because we assume that "solution search" has much to do with new requirements and therefore new information is required. Figure 5.10 depicts the relative amount of the transfer of new information in the different types of critical situations. In situations with a positive solution search (+), the designers produce the highest amount of new information, whereas in situations with a negative solution search (–), designers produce a significantly lower percentage of new information. In situations of negative solution analysis and decisions, we found an equally high percentage of new information transmitted; obviously, the communication of new information during later phases of analysis and decisions on solutions is not very useful.

Furthermore, interesting differences can be noticed concerning the category "searching for information." This category consists of two subcategories,

Table 5.4. Excerpt of a categorized dialogue in a positive critical situation of the type solution search

Time	Person	Transliteration	Cat
06:33:46	B	Okay – that means that you'll take a set-screw – thereupon?!	ip
06:33:48	C	A log next to it and an adjustment-screw.	ip
06:33:58	C	The problem is that I cannot screw it tight to fix it.	nk
06:34:00	B	What can't you fix?	aa
06:34:01	C	This angle, I do not have any space left.	if
06:34:04	B	On this other part?	aa
06:34:18	C	Yes, you see, basically I did create a little basis: this is a vertical slot and there I will make a stud.	ip
06:34:24	B	Hmm.	r
	C	Don't say Hmm". What shall I do otherwise?	na
06:34:28	C	Or perhaps a flathead-screw, one could weld it on, you are right. In any case, I need a spike.	pk
06:34:34	B	Yes, that means, you have to unfix that …	vk
06:34:48	C	(While outlining) Yes, here is now then … You have this … this here is this long, this slot, in which the whole can be pushed up and down. That here is only a through-hole.	if
06:34:54	B	Just let me ask you: what is that for a welded joint? Which part are you welding to which other part?	aa
06:35:20	C	Oh no, I see, stop … ahh, in the lower plate. I mean this one, there is this big notch, you see, and the plates are overlapping here. I did insert such a piece that has in its height this cutting …	if
06:35:24	C	… so that this is always overlapped and sealed up.	ip
06:35:27	B	And the part moves in the way you are moving the things.	ip
06:35:31	C	Right. If I push them together, then this one is always moving with it.	if
06:35:34	B	That means that backwards there is once again somewhere such a …	if
06:35:44	C	Right, there is once more such a shape, so that I also can move here to and from, okay? In such a way that this always overlaps and seals up.	pk
06:35:47	B	Now, how do we get that …?	aa
06:36:01	C	The only thing is now the …, that I had to screw it here, too. Okay, then, I think, if I could do it with three screws and put my bearing here, whatever bearing, and I make here a screw with a slot …	vk
06:36:10	C	And here below two – that looks ugly, doesn't it?	nk
06:36:17	B	How much space do you have for this angle related to the height? For example, there below is not much …	aa
06:36:23	C	For example, and at the head there is also not much space – you see that?	ab
06:36:31	B	Yes, therefore for the first it would be perfect, if you would screw them on this axis. So that you there …	in
06:36:39	C	With two screws?	aa
06:36:43	B	Yes, at first you try with two screws – like the flange-bearings, the Y-flange-bearings, they have two screws. So …	ip

Table 5.4. continued

Time	Person	Transliteration	Cat
06:37:00	C	That would mean . . . yes, I also thought about this solution. You see here, I have drawn here that I have to weld on and that I make the plate so, and then here a slot and there a slot.	vk
06:37:22	B	And if you disengage from that silly journal-bearing and reconsider to take still another one?	nk
06:37:38	B	Building a box. That here would be the box, that is the base-plate [he outlines his idea on a piece of paper]. This corresponds to that, okay? And then you can make a box and you make a lid for the box and then you'll have the box.	in
06:37:56	B	But here at the top and below is open, okay? And here into you'll set your flange-bearing. And if the flange-bearing would go beyond here, then you can even set it here through and interrupt this thing here.	ip
06:37:59	C	Aaahhh, did you consider that the flange bearing would build the whole thing bigger?	nk
06:38:00	B	Yes, that is no problem.	pz
06:38:04	C	No, it is a problem. I have no space here at the top and I just can't place the box.	nk
06:38:18	B	So, now there is the flange-bearing in here. There are the screw-holes – and there you make your two things, and there you make an opening . . .	in
06:38:24	C	Do you mean these two-? Bearing?	aa
06:38:40	B	Yes, for example – a two-hole.	pz
	B	And here behind you make the same stuff, and in case of need the same fraction and the same holes. And then you can screw here and there.	ip
06:39:10	B	And then you still need slots. And these here, these are slots. Hm? Perhaps in case of need you can make the fraction two times, then you have here the slot once, and here you can push it up and down, and here lays the Y-bearing . . .	ip
06:39:35	B	Of course you have to invent something to bring the bearings there in. So that you can build them in, but I think that will be possible.	vk
06:40:06	B	For example, you could take a round one – perhaps they build a little bit smaller than these two ears. Or even a rectangular.	ip

questions of all sorts (e.g., ". . . What can't you fix?") and demands for attention (e.g., ". . . and at the head there is not much space either – do you see that?").

Unexpectedly, both subcategories of information search (focused questions that can be answered directly, as well as demands for attention, i.e., asking in a way that demands reflection on an issue) occur significantly less often in positive than in negative situations of solution search. Does this mean that asking questions could lower the quality of solution search and thus lead to less new information? Obviously it is more important to "give space" to a colleague and encourage him or her to talk about proposed solutions than to ask questions, as we can conclude from a comparison between Figures 5.10 and 5.11.

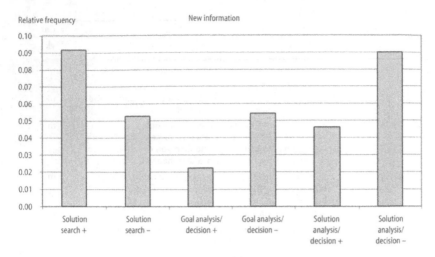

Figure 5.10 Relative frequency of new information transmitted in different types of critical situations.

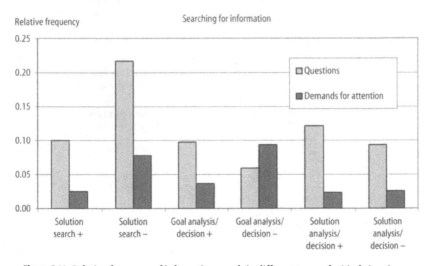

Figure 5.11 Relative frequency of information search in different types of critical situations.

The differentiation into types of information search (focused questions or demands for attention) does not seem to elicit a deeper understanding of successful communication in the solution search. So we still ask what type of information is most important in encouraging successful communication and in preventing less successful communication in situations of solution search. In what follows we focus on the role of positive and negative statements in critical situations. Figure 5.12 shows how often negative evaluations ("generally negative" and "negative design-related" statements) occurred in

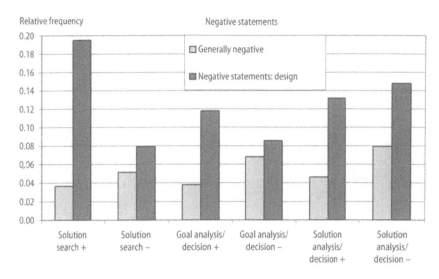

Figure 5.12 Relative frequency of negative statements in different types of critical situations.

successful and less successful critical situations. A generally negative statement bears no relation to a feature of the discussed design – for example: "This is all useless under these conditions." Contrary to these unspecified evaluations, a negative design-related statement is, for example: "The problem is that I cannot screw it tight to fix it."

Our research shows that crucial design aspects are discussed significantly more frequently in successful situations of solution search than in less successful ones. The most interesting finding is that the number of negative design-related statements in positive solution search situations (+) is very high, whereas the number of general rejections (negative items) is very low in these successful situations.

Contrary to the emphasis on content regarding negative statements, positive statements are mostly general and not design-related. (e.g., "Yes, that is no problem"). Figure 5.13 shows that in positive solution search situations (+), many more general positive statements and fewer design-specific positive arguments occur. An example of a design-specific positive statement is "Or perhaps a flathead-screw, one could weld it on, you are right." In negative solution search situations (–), we find only general positive statements and no design-related positive arguments.

A chi^2 test shows that all the reported differences are highly significant (p < 0.01).

How can we interpret these findings? The analysis above indicates, on the one hand, the importance of criticism on the basis of professional design knowledge, but, on the other hand, it points to the group climate as an additional ingredient in a successful, creative solution search: the amount of general positive statements in group communication that contributes to a positive climate balances the design-related negative arguments and makes them acceptable and useful.

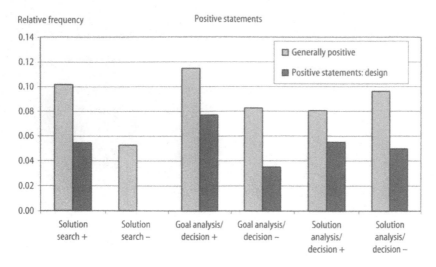

Figure 5.13 Relative frequency of positive statements in different types of critical situations.

What are the factors supporting communication in successful solution search situations?

To interpret the observations concerning design representations in critical situations, we must take into account the factors of the individual, the group, and the external conditions compiled in this study as described in section 2 of this chapter. Indeed, we find a good group climate to be a major influencing factor supporting communication in successful situations of solution search, as Figure 5.14 demonstrates. This figure is based on the analysis of 28 situations of successful solution search, in which factors influencing the quality of the solution search were identified. For every situation we made an annotated chart of the causal relations between influencing factors related to the task to individual prerequisites, to prerequisites of the group and to the external conditions, and parameters of the design process (cf. section 3.1). By summing up these single charts, we gain an insight into the mechanisms that have led to 28 situations of successful solution search. These mechanisms are diagrammed in Figure 5.14. The thickness of the arrows is proportional to the frequency (percentage) of the relations found in this type of critical situation. The thickness of the box frames is proportional to the frequency (percentage) of the factors identified in all critical situations of "successful solution search." An example of a successful solution search situation is given in the dialogue in Table 5.4.

The results obtained from the observation of 28 successful solution search situations reveal that the generation of ideas depends largely on the availability of information concerning the requirements and knowledge of possible solution principles. In 75% of all positive solution search situations, a good group climate was revealed as an important group prerequisite, and in most cases (54%) the good group climate triggered communication, which in itself was an important prerequisite of information availability. There are,

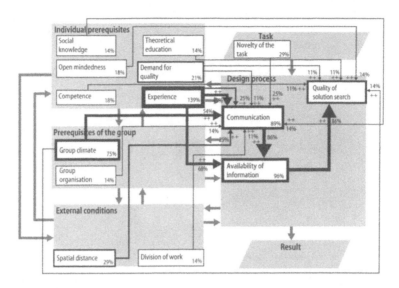

Figure 5.14 Mechanisms leading to a successful search for solutions (+) in 28 critical situations of this type. The figure 139% in the box "experience" is caused by the fact that high experience in one critical situation may serve both as a source for communication as well as a cause for the availability of information, so that we have to take "experience" into account twice in one critical situation. Thus, a percentage of over 100% can only result if one factor is related to more than one other factor.

of course, many other aspects that contribute to the quality of a solution search, but regarding the group, "group climate" is more important than has been assumed in recent studies undertaken in the context of social psychology (Witte and Lecher 1998).

Another important group-related factor supporting the main design representation "communication" is group organization. In situations of positive solution search, however, group organization proved to be less relevant and occurred in only 14% of all situations of positive solution search. (For example, the informal discussion between designer C and designer B in Table 5.4 did not take place because of group organization; rather, it was based on the trustful relationship between the colleagues C and B.)

Achieving adequate group organization, together with good group climate, is among the major responsibilities of the group leader. The optimization of important external conditions, such as spatial distance between colleagues and the division of work, is more or less the responsibility of the management. This makes the quality of leadership a major prerequisite in the background of successful communication in solution search situations, and thus group creativity. But we have to consider that the individual designer with his or her experience and knowledge has an impact, too: information transfer is mostly based on the experience of designers. Moreover, the open-mindedness of the individual designer, a readiness to accept new ideas and a high demand for quality, leads to an intensive search for solutions (see also Badke-Schaub and Buerschaper 2000). These findings are corroborated by other studies (Perkins 1988; Sternberg 1988) that have shown that domain-specific knowledge is not sufficient for successful problem solving because the adaptation

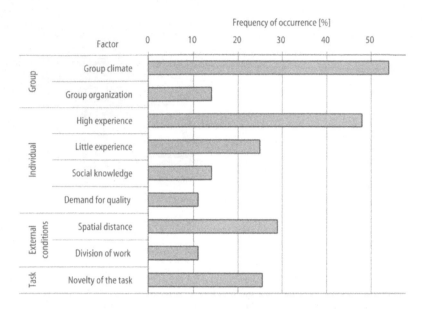

Figure 5.15 Factors influencing communication in situations of successful solution search.

and restructuring of knowledge is also related to personality characteristics such as open-mindedness and tolerance for ambiguity.

Figure 5.15 illustrates how frequently different factors support verbal information transfer in critical situations of successful solution search.

Conclusions

Thinking about external design representations we instantly associate design representations with artefacts such as sketches, drawings, or models. But whereas sketches are a basic design representation for every routine work in product development, critical situations in the design process underlie other representational conditions. In critical situations communication between colleagues is the most important design representation: engineering designers contact colleagues in nearly 90% of the critical situations identified along the design process. Therefore, the appropriateness of communication in different types of critical design situations turns out to be an important prerequisite for successful design work. Whereas routine work can be supported adequately and quickly by computer-based information tools, social information transfer in critical situations must be emphasized and supported.

Usually, designers prefer to search for information by asking their colleagues because, in doing so, they receive much additional information, including a rough sense of the quality of their own idea, solution, or decision. The study suggests that it is important for group members to know the rules and mechanism to enforce adequate information transfer during these situations. We know that it is important to encourage designers (and all people, for that matter) to preserve the individual's feeling of competence. Negative evaluations and too many questions may contradict this need and thus break the social rule that requires that the group be friendly and cooperative.

Nevertheless, the development of solutions, as well as positive group performance in general, is hardly based on positive affirmations; rather, it is based on critical evaluations. Furthermore, a good group climate does not guarantee a good group product, although it seems to be an important setting for it. As our study illustrates, it is necessary to state the negative aspects of evaluations in the particular design context that applies to it. Generally, negative statements do not inspire a successful solution search, whereas positive statements may be important in stabilizing the actual situation and the group atmosphere. A productive discussion may evoke and encourage the communicant to create, revise and modify ideas, thoughts and hypotheses. In "creative" moments, such as solution search situations in particular, a good group climate seems to be the basis for an open discussion of different design aspects, especially crucial aspects of the solution. This finding makes the group climate an important issue of design communication and representation in order to create a high performance design team.

Note

1. This method of "critical situations" sounds similar to the "critical incidents" of Flanagan (1954) or the "critical moves" of Goldschmidt (1996), but it follows another concept because the identification of the critical situations takes place according to the requirements of the design process.

References

Badke-Schaub, P 1998. Determinanten der Informationsverfügbarkeit von Arbeits gruppen in der Praxis. In Vortrag auf dem 41.Kongreß der Deutschen Gesellschaft für Psychologie: Dresden.

Badke-Schaub P and C Buershaper 2000. Creativity and complex problem solving in the social context. In Decision making: Social and creative dimensions, edited by C.M. Allwood and M. Selart. Dordrecht: Kluwer.

Badke-Schaub P and E Frankenberger 1998. Zwischen Aufwand und Erkenntnis: Zur Aussagekraft von strukturierten Interviews über zentrale Mechanismen in der Konstruktion. Memorandum Nr. 29, Institut für Theoretische Psychologie, Univeristy of Bamberg.

—— 1999. Analysis of design projects. Design Studies 20:481–494.

Badke-Schaub, P, J Stempfle, and S Wallmeier 2001. Transfer of experience in critical design situations. In Design management: Process and information issues, edited by S Cully, A Duffy, C McMahon, and K Wallace. London: Professional Engineering Publishing, pp 251–258.

Badke-Schaub, P and T Tisdale. 1995. Die Erforschung menschlichen Handelns in komplexen Situationen. In Computersimulierte Szenarien in der Personalarbeit, edited by B Strauß and M Kleinmann. Gottingen: Verlag für angewandte Psychologie.

Blessing, L 1994. A process-based approach to computer-supported engineering design. Thesis, University of Twente, Enschede, the Netherlands. Cambridge: Black Bear Press.

Cross, N and A Cross 1996. Observations of teamwork and social processes in design. In Analysing design activity, edited by N Cross, H Christiaans, and K Dorst. Chichester: John Wiley & Sons.

Dörner, D 1998. Thought and design - research strategies, single-case approach and methods of validation. In Designers: The key to successful product development, edited by E Frankenberger, P Badke-Schaub, and H Birkhofer. London: Springer, pp 3–11.

Dörner, D and AJ Wearing 1995. Complex problem solving: Toward a (computer simulated) theory. In Complex problem solving: The European perspective, edited by PA Frensch and J Funke. Hillsdale, NJ: Lawrence Erlbaum Associates.

Ehrlenspiel, K 1995. Integrierte Produktentwicklung. Methoden für Prozeßorganisation, Produkterstellung und Konstruktion. München: Hanser.

Eisentraut, R 91997). Styles of problem solving and their importance in mechanical engineering design. In: Engineering psychology and cognitive ergonomics, vol. 2, edited by D. Harris. Aldershot: Ashgate, pp. 363–370.

Flanagan, JC 1954. The critical incident technique. Psychological Bulletin 51:327–358

Frankenberger, E 1997. Arbeitsteilige Produktentwicklung – Empirische Untersuchung und Empfehlungen zur Gruppenarbeit in der Konstruktion. Düsseldorf: VDI-Verlag.

Frankenberger, E and P Auer 1996. Standardized observation of teamwork in design. Research in Engineering Design 9:1–9.

Frankenberger, E and P Badke-Schaub 2000. Kritische Situationen als Zugang zum Problemlösen in der Produktentwicklung. In Konstruieren zwischen Kunst und Wissenschaft edited by G Banse and K Friedrich. Berlin: Edition Sigma Rainer Bohn Verlag, pp 237–260.

Goldschmidt, G 1996. The designer as a team of one. In Analysing design activity, edited N Cross, H Christiaans, and K Dorst. Chichester: John Whiley & Sons, Chichester.

Görner, R 1994. Zur psychologischen Analyse von Konstrukteur- und Entwurfstätigkeiten. In Die Handlungsregulationstheorie: Von der Praxis einer Theorie, edited by B Bergmann and P Richter. Göttingen: Hogrefe, pp 233–241.

Hacker, W, P Sachse, and F Schroda 1998. Design thinking – possible ways to successful solutions in product development. In Designers: The key to successful product development, edited by E Frankenberger, P Badke-Schaub, and H Birkhofer. London: Springer, pp 205–216.

Hales, C 1987. Analysis of the engineering process in an industrial context. Dissertation, Cambridge University.

Klimoski, R and S Mohammed 1994. Team mental model: Construct or metaphor? Journal of Management 20:403–437.

Pahl, G and W Beitz 1997. Konstruktionslehre. Handbuch für Studium und Praxis. 4th ed. Berlin: Springer, Berlin.

Perkins, DN 1988. Creativity and the quest for mechanism. In The psychology of human thought, edited by RJ Sternberg and EE Smith. Cambridge: Cambridge University Press.

Pugh, S 1990. Total design; integrated methods for successful product engineering. Reading: Addison-Wesley.

Städel, T 1987. Problemlösen, Emotionen und Kompetenz. Die Überprüfung eines integrativen Konstrukts. Roderer: Regensburg.

Sternberg, RJ, ed. 1988. The nature of creativity: Contemporary psychological perspectives. Cambridge: Cambridge University Press.

Witte, EH and S Lechner 1998. Beurteilungskriterien für aufgabenorientierte Gruppen. Gruppendynamik 29:313–325.

6

Impromptu Prototyping and Artefacting: Representing Design Ideas through Things at Hand, Actions, and Talk

Gilbert D. Logan and David F. Radcliffe

Introduction

Design offices have always been littered with physical artefacts, including components, prototypes and models, as well as drawings and various paper-based design objects. Radcliffe and Harrison (1994) noted that physical artefacts, including existing products and prototypes, are an important part of the design environment in a small manufacturing company. Product development involves numerous movements across the boundary between physical artefacts and abstract representations, including drawings, sketches and lists. Notwithstanding the evolution of the CAD (computer-aided design) systems and the promise of virtual design tools (e.g., Krovi et al. 1997; Furlong 1997), physical objects continue to have a place in the contemporary design office. However, relatively few empirical studies have focused on the role of artefacts in engineering design.

Tang (1991) found that during group sketching activity, designers used drawing, lists, and gesture to store information, express ideas and mediate interaction. These actions were supported by talk, although Tang treated this as a substrate to the design actions and did not examine the role of talk in respect of the actions. A subsequent study by Radcliffe and Slattery (1993) of the work of a cross-discipline, rehabilitation engineering team observed that they used lists, gesture, mimicry, and physical interventions to express and test design ideas, to explore the context of the task, to mediate interaction, to negotiate closure and to store information. Harrison and Minneman (1996) observed in the Delft Protocol that artefacts serve a variety of functions in the hands of designers. They concluded that "the processes of interaction with objects have communicative value and alter the dynamics in multi-designer settings."

Analyzing the same group of designers in the Delft Protocol, Radcliffe (1996) noted that, on several occasions where objects were recruited to demonstrate a design proposal, team members augmented their actions with sounds to accentuate the meaning. For example, the sound of spot-welding was used as part of a presentation to convey to team members how an assembly would be made. While no formal analysis was conducted into the relationship between engagement with artefacts and talk, Harrison and Minneman (1996) noted that, as the body of referents increased, the use of pronouns in designers' speech increased. Thus, talk and action with artefacts are intertwined in ways that have not been explored fully.

This chapter presents a fine-grained, empirical study (Logan 1999) of the relationships between design actions, physical objects, and talk in design discussions in a cross-discipline team. We use the term "artefacting" to describe the combination of utterance and physical interaction used to communicate complex messages that have design implications. Physical interactions include gestures, interventions with hands or objects, mimicry in which the body simulates ideas, and the use of physical artefacts at hand during a design act. Some of these artefact interactions involve the construction of artefact assemblies – prototypes through which participants provide physical representation of their ideas and also gain immediate experience of the prototype in the current physical environment. This impromptu prototyping was recognized by Horton (1997) as having the potential to develop a designer's "device in mind" into an overall design intent.

This study is based on a cross-discipline, rehabilitation engineering[1] team at work in a seating clinic. Rehabilitation engineering is often provided as a service to people with a disability in hospital, university, or community health settings to evaluate, prescribe, devise, and provide assistive devices to increase their independence and reduce handicap. A Rehabilitation Engineering Centre (REC) employs rehabilitation (professional) engineers, technicians, physiotherapists and occupational therapists who work in a team to offer unique combinations of knowledge and skill to help solve physical problems experienced by a client. A rehabilitation engineering team usually practices face-to-face with the client and the client's care-providers to: (1) acquire information about the client's physical situation, functional ability, and performance of current assistive devices such as a wheelchair; (2) experiment to understand how functional performance, postural control, etc. is enhanced or exacerbated by various interventions, and (3) devise unique solutions to problems based on the outcome of intervention and the trialling of various items of equipment. Much of the information on which decisions are based is acquired by observation. The cross-discipline team in this study specializes in aiding clients with severe physical disability who experience difficulty with mobility, comfort and control of sitting posture, and pressure sores. Figure 6.1 illustrates a seating clinic in progress. The team undertakes an assessment of the client's needs, then designs and manufactures custom seating and related devices to ameliorate specific problems. The assessment is conducted according to a set of questions and headings that seek to probe for specific and general information. The assessment aims to encourage discussion and experimentation with the client by trying various seating

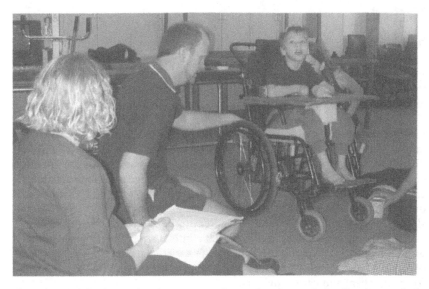

Figure 6.1 Rehabilitation engineering team members gathered around the client in a typical seating clinic.

postures in association with artefacts such as commercially available seating aids and objects at hand, off-cuts of polyurethane foam, plywood and anything that offers some potential to be useful to assist designing.

Each member of the team brings particular expectations about what needs to be accomplished in the seating clinic. Their understanding of the process is influenced by their professional training. The engineer understands the processes as one of design and manufacture invested with all the connotations that these terms have for an engineer. Team members might expect to propose design concepts and to document them in the form of drawings, or at least sketches and associated lists of relevant information which is sufficient to manufacture the seat. On the other hand, a therapist might approach the task from the perspective of diagnosis and prescription. The therapist's expectation is of an investigation with limited amounts of documentation of the prescription. The technician is likely to be focused on the practicalities of making a seating system.

This contrasting set of traditions and the consequential lack of common ground in how they do their work, makes the rehabilitation engineering team a very interesting one to study. The rehabilitation engineering team relies heavily on talk and physical interactions or artefacting for communication. Information is gathered intermittently throughout the clinic session and recorded separately by the different professionals involved. There is no commonly agreed method for systematically analyzing this information when generating a design for the seating system. This behaviour is similar to what might be expected of any cross-discipline team that does not share a common heritage of work patterns and means of documentation. Thus, the results of the study presented in this chapter may be generalized, with caution, to other product development settings that involve cross-discipline teams.

Methods

Data gathering

Video provides a means to record accurate detail of human interaction and the capability to scrutinize repeatedly the detail for social action and activity. Numerous studies of naturalistic human activity have sourced data for analysis from video recordings of the activity (Goodwin and Goodwin 1996; Kleifgen and Frenz-Belkin 1997; Heath 1997). The cross-discipline, rehabilitation engineering team was videotaped performing assessments of clients, during which information is gathered about the client and his or her specific equipment issues. The investigation of potential design solutions occurs by experimentation and considerable discussion (prototyping of customized devices occurs concurrently with the experimentation in the adjacent workshop). Assessment involved examining existing equipment, asking questions, and discussing experiences to elicit specific information about the client that was relevant to developing the appropriate assistive devices. A prepared set of forms guided the assessment by prompting for the information to be sought from the client and care-providers.

Video recordings were made using a video camera with a wide-angle lens mounted high in the corner of the assessment room and focused on the location where the participants tended to group. A flat plate microphone fixed to a wall provided the sound input to the video camera. Clients and care-providers were asked in advance if they would participate in the video recording and signed informed consent documents. All participants were made aware of the presence of the video camera before the assessment commenced. No further reference to the video camera was made during an assessment session. Video recordings were made of 24 assessment sessions, selected randomly, during a three-year period. Work practices did change during that time due to the normal events that occur in a work place. Staff reflected on work practice and implemented changes to improve the process, such as the prompts in the client files. Staff changes occurred with therapy staff rotating through this clinic; however, the rehabilitation engineer and technician were constant team members across the recordings. No intervention was attempted to limit change or to control practice. While this may be seen to limit the reliability and validity of the data (Peräkylä 1997), this was a study of the practice of a rehabilitation engineering team, and the normal processes that affect a work group prevailed. Each client and care-provider group was different and presented with different issues to be resolved and there was large variation in the amount of investigation performed and the development of design solutions.

After the first year of the project, a small video camera was made available to the team members to use if they wished. Team members had experienced direct benefit from video recording as they found specific information that had not been documented. Video was also embraced by team members as a useful tool for recording specific information about clients. The hand-held camera close-up data supplemented the wall-mounted camera recordings that recorded the whole workgroup and showed most of the detail (verbal and non-verbal behaviour) of the participants. Occasionally, data was lost because the view-field was obscured by a participant moving between camera and the current activity, or a participant chose to work at the far side of the wheel-chair which was obscured from the camera.

Data analysis

It was observed in each of the 24 videotapes that participants' communications concerning the work at hand frequently contained talking in association with actions. Sometimes participants made actions without talking. Seven of the videotapes with extensive segments of talking and actions were selected for detailed analysis. The total length of the segments in the seven videotapes was 2 hours 45 minutes. Data was extracted from the videotapes chronologically by transcribing the participants' talk,[2] describing the actions[3] and noting the artefacts used, and timing the length of the talking and acting. Table 6.1 is an example of the action description style adopted to accompany the transcribing of talk.

Categorizing action

In the videotape analyzed, many artefacts were used and many actions were applied to them across a wide range of talk-type. None the less, common and recurrent actions were easily recognizable. Categories to code participants'

Table 6.1. Description of action adopted in transcribing videotapes

Time	Speaker	Transcription
1:00:40	Engineer (Engr)	He has been going into that sort of posture [ART Engr moves pelvis into sideways-sitting posture]* all the time which you know encourages that sort of curve [ART traces out curved line in the air with the spanner he is holding]. (The action description was written into the transcript, encased in square brackets [ART] with the start of action synchronized with the talk accompanying it.)
1:00:52	Occupational Therapist (Occ. Th.)	I'll put his arm up on the armrest [Occ. Th. walks over and grabs hold of Stephen's left arm at wrist and elbow and lifts it up off the left side seat and places it on the left armrest, straightening Stephen's trunk in the process.]
1:01:00	Engineer	That is something too he has got no where to he can't keep that arm on anything to help {Occ. Th.: the arm allows him to sit back}.* The armrests don't go back enough [ART Engr moves right arm in a sweep towards front of the wheelchair towards the back]. {Occ. Th.: Does that feel more comfortable now, Stephen?} They are not padded big enough. You know, if he had something over that side [ART Engr has outstretched hand pointing over the side of the wheelchair] to support his arm in this sort of position [ART Engr extends same arm and points in general direction of Stephen's supported left arm which is now resting on the left armrest], then that would help. {Orthotist: Uhm.}

*[. . .] Square brackets indicate a description of artefacting that accompanied the person's talking. The person performing the artefacting is labelled, e.g., Engr – Engineer.
*{. . .} Curly brackets indicate that the labelled person, e.g., Occ. Th. – Occupational Therapist – interjected during or briefly interrupted the main speaker's talk.

actions used while communicating seating design information were developed based on evidence, visible or audible in observation of the video tapes, as proposed by Tartar (1989). Harrison and Minneman (1996) studied the use of objects by a team of three designers during a videotaped design activity and devised a five-category system for the use of objects in design. This system was expanded to six categories and modified to cover activities observed in the rehabilitation engineering videotapes that Harrison and Minneman did not consider. The six action categories and their indicative types of action shown in Table 6.2 offered an objective classification scheme that could be applied by observing the participants' actions accompanying their talk, or occurring in silence. The types of action listed in the right-hand column of Table 6.2 were developed by generalizing into simple, unambiguous movements and actions all the actions observed across the tapes. Photographs showing some examples of the Action category appear in Figure 6.2.

Categorizing talk

A system to categorize talking arose from evaluating videotape transcripts and reviewing the videotapes to find obvious groupings. The development of both the action categorization and the talk categorization schemas was assisted by members of an engineering research practice group to refine the groups and the application of coding rules. Table 6.3 shows the final schema that was developed to code participants' talk.

The consistency of the categories that were chosen for coding the talk and action data was tested by comparing the coding that was applied by three test coders and the coding that the authors had applied. One test coder was familiar with the work at the Seating Clinic; the other test coders had had no exposure to the Clinic. Each coder was: (1) given an introduction to the task they were to perform; (2) provided with notes on how to code videotape, event[4] by event; (3) supplied a videotape of two clips of the Seating Clinic video lasting 20 seconds and 2 minutes, and (4) given rating sheets with the transcript of each event on the videotape segments and space to mark his or her choice

Table 6.2. Categories developed to classify actions that participants use associated with designing

Action category	Types of action observed
Constructing	Mock-up using artefacts, hand shapes, animations.
Locating/Indicating	Touch, tap, trace around, scribe marks on an object.
Measuring	Application of conventional measuring instrument, plus approximations using hand-to-hand span, hand-to-object distance, thumb-finger distance, etc.
Demonstrating function	Using an artefact (object) by applying force, (push, pull, rotation) to demonstrate some feature or a desired effect.
Examining	A participant investigates an object by himself.
Gesturing	Look at, glance towards an object, person. Pointing to an object but not touching it. Waving an arm at an object or person.

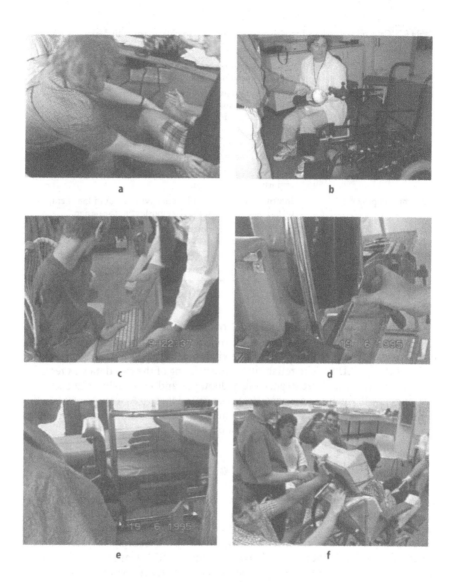

Figure 6.2a–f Examples of some Action categories selected from videotapes.
a Demonstrating function: therapist applies force to the leg of another therapist to illustrate how the client's leg should be positioned by the seating.
b Gesture: engineer points to the component held by the orthotist as he responds to a question asked by the client (out of the picture).
c Measuring: engineer uses his open hands to indicate a width of the easel to support the laptop computer.
d Measuring and locating: engineer illustrates proposed location and size of the component to secure an oxygen cylinder to the frame of a wheelchair.
e Constructing: technicians use some wood blocks and a hand to mock up an oxygen cylinder support in a wheelchair.
f *Multiple acting*. Constructing: orthotist (foreground) mocks up a strap attachment for a chest harness. Locating/indicating: physiotherapist (standing on the left) traces proposed trim line for the headrest with finger. Demonstrating a feature: client's father (sitting, at top) illustrates how a chest harness strap was positioned in an earlier design of seating system.

Table 6.3. Categories developed to classify the talk that participants use with design

Talk category	Defining feature
Design proposal	Contains data and concepts to specify a physical requirement to be constructed.
Dimension/location	An actual dimension or estimation of size, position, orientation of something that has design relevance and may be included in the completed construction.
Explanation	A response to a question or statement by another participant that provides additional information or should enhance understanding.
Information client	Information and data about the client's body that is important to the design.
Information physical	Information about the physical environment of the client, including equipment that is important to the design.
Comment	An unsolicited utterance that may provide additional information or be an opinion, or observation.
Question	An indication that something is unknown and should be known.

of classification of talk and action, and other details in each event. The correspondence of the test coders with the author ranged from 51% to 83%, with an average of 66%. The coding of all the videotape data was performed by one of the authors (G.D.L). The reliability of the refining of the raw data was tested by retranscribing and recategorizing videotapes and comparing the second pass check with the first pass records. It was found that 10% of events required some reassignment of a data entry. The talk and action data that was extracted from the videotapes was stored in a relational database.

A Sample Exploratory Analysis of a Videotape

Exploratory analyses of videotapes were conducted to provide an insight into talk and action. Each videotape was reviewed by précising sequentially the major occurrences, which involve artefacts and action or provide significant information that was relevant to design during the Seating Clinic. It was observed that the introduction of artefacts into the REC team usually started an investigation or a series of question–answer interactions by participants that revealed useful data, understanding and ideas. An example of a design activity segment analysis from a videotape is presented in the following section.

Nicholas' "final design" segment

Nicholas is a teenage boy born with congenital amputations of his arms and his right leg. His left leg is an incomplete limb, resembling a flipper, that can be moved and controlled, but which is incapable of load bearing. Nicholas can move independently by bottom shuffling. He also uses a power wheelchair that has powered vertical movement of the wheelchair seat. Nicholas can shuffle in and out at floor level and raise the seat to a convenient height

Table 6.4. Occurrences identified in segment 8 of videotape "Nicholas"

Tape Time	Occurrence
20:37	Mother takes the wedge from under the computer and places it on the left rear of the netting backrest after Nicholas asks for some extra support in the backrest.
20:56	Question and answer about a "tight" muscle in Nicholas' back.
21:52	Engineer decides to make the backrest for the wheelchair, then copy it for the computer.
22:10	Engineer establishes the angle Nicholas wants the computer tilted at.
22:22	Nicholas requests his old backrest to lean against which mother sets up for him.
23:00	Nicholas says he likes the support of this backrest but it is hard to type.
23:17	Physio shows that the lateral wings get in Nicholas' way. Nicholas says no, he likes their side support.
23:50	Engineer suggests trying a modular-style backrest. Uses tape measure to check Nicholas' trunk width to select backrest size.
24:00	Engineer measures dimensions of the assembled prototype and talks about base size.
24:41	Modular backrest replaces the unit Nicholas has been resting against.
24:54	Nicholas requests that backrest be on an "X" and rotates his trunk around vertical axis.
25:02	Engr mimics Nicholas' movement with a backrest he is holding.
25:22	Physio suggests that it is on a spring so it won't just flip.
25:54	Discussion about the shape of the backrest.
26:55	Engineer measures the angle of foam wedge.
28:00	Backrest is tried for fit in the wheelchair.
28:23	Decisions made about how to proceed with manufacture.

for driving and working on tasks. The REC team is investigating seating to assist Nicholas to access his laptop computer (Figure 6.3). This analysis is of the final segment of eight design segments in the Nicholas videotape. In the previous segments basic parameters have been established for the seating. As Table 6.4 shows, this segment contains the following occurrences; unexpected design ideas are revealed.

Commentary on Nicholas, segment 8

Segment 8 contains significant artefact and talk and action features. The flexibility of the simple artefacts is demonstrated by the multiple uses of the wedge section polyurethane foam block, first as an angle support under the laptop computer, then as extra support behind Nicholas' back. The latter use of the foam block starts a line of enquiry, providing new information about Nicholas' back support needs. Testing another type of backrest provides data on the shape that Nicholas considers offers good support. While using his computer in an assembly of artefacts mocking up a proposed seat, Nicholas suddenly executes some trunk twists and suggests of the backrest, "Wouldn't that be good if it was on like a like a like an X sort of thing like it can like tilt." Later the engineer is able to verify the backrest motion that Nicholas

Figure 6.3 Client demonstrating a new feature of a backrest: "Wouldn't that be good if it was on like a like a like an X sort of thing like it can like tilt." Engineer in the background holding a sample backrest immediately uses the artefact to feedback forward and back rotational movements around horizontal and vertical centrelines of the backrest to the client to test if the information has been interpreted properly.

desires by mimicking his movements using a backrest he is holding. This segment illustrates the value of artefacts in assisting design through (1) demonstrating, testing, and verifying physical information relating to shape, size, and orientation of design elements; (2) introducing a new feature or idea to encourage physical or observational participation and interaction; and (3) inducing a state of curiosity in participants, so encouraging questions, answers, comments and experience-based ideas and opinion.

Results

The total number of events recorded was 882. This comprised 319 events in which only talk was used, 41 events in which action occurred without talk and 522 events combining action with talk, hereafter labelled Talk & Action. Nearly 60% of the communication work was performed using Talk & Action. In events in which the Talk was developing or relating a design idea, the use of Talk & Action rose to 86%. Table 6.5 lists the number of events that contained Talk, or Talk & Action according to the talk-type.

The occurrence of each action-type for the events containing action (i.e., Talk & Action and Action performed in silence events) is shown in Table 6.6. Sequences containing two or more actions performed by a speaker occurred in 40.5% (228) of the Talk & Action events. Actions in sequence were separable as discrete entities. Actions in sequence often showed a smooth flow from one to the next. Some 59.5% of events had a single action and 84% of events had one or two actions. The number of events with more than three

Table 6.5. The number of events of each talk-type

Talk-type	Number of events	Percentage
Design proposal	161	19.2%
Dimension/location	53	6.3%
Explanation	97	11.5%
Information-physical	92	10.9%
Information-client	42	5.0%
Comment	222	26.4%
Question	174	20.7%
Total	841	100%

Table 6.6. Count of action-types in the data

Action-type	Number of occurrences of the action-type	Percentage of all action-types	No. of events in which the action-type was observed
Constructing	131 (121)*	14.2%	92
Locating/indicating	277 (273)	29.9%	191
Measuring	92 (82)	10.0%	79
Demonstrating function	125 (118)	13.5%	105
Examining	148 (137)	16.0%	133
Gesturing	153 (153)	16.4%	117
Total	926 (884)	100%	

* Amounts in brackets do not include action performed in silence events, i.e., action without talk.

actions was small. In most events (93%), speakers performed actions during their talk. The amount of action performed before or after talk was small – 4.5% and 2.5% respectively.

Table 6.7 shows the number of action-types that occurred for each talk-type in the videotapes analyzed. The top number in the data cells is the total of the action-type for the corresponding talk-type in that row. The bottom number (in brackets) in a cell, termed the "action value," is the normalized amount of action per event of talk-type in the row. It offers a comparison of action across talk-types with different totals of events.

The distribution of cells shows the following characteristics with respect to the average "action value":

1. The "Locating/indicating" column contains "action values" two to three times the table's average value for talk-type "Design proposal," "Dimension/location," and "Explanation," and between 1.08 and 1.7 times the average for the remaining talk-types. No other action-type involves such an extensive use in talk.

2. "Locating/indicating" and "Constructing" action-types were used frequently with talk-type "Design proposal." The "action value" for these action associations is ranked highest (0.759) and second highest (0.635) in the whole Table. "Design proposals" are associated with a below-average use of the action-types "Measuring" (0.146 "action values"), "Demonstrating function" (0.183), "Examining" (0.16), "Gesturing" (0.248). Of these four, "Gesturing" actions were used most.

Table 6.7. Action-types used with talk in Talk & Action events in the videotapes

Talk-type	Action-type					
	Construct-ing	Locating/ indicating	Measuring	Demon-strating function	Examin-ing	Gestur-ing
Design proposal No. of events: 137 No. of actions: 293	86 (0.635)	104 (0.759)	20 (0.146)	25 (0.183)	22 (0.16)	36 (0.248)
Dimension/location No. of events: 45 No of actions: 74	5 (0.111)	25 (0.556)	28 (0.622)	2 (0.044)	10 (0.222)	4 (0.088)
Explanation No. of events: 73 No. of actions: 132	7 (0.096)	46 (0.631)	13 (0.178)	21 (0.288)	21 (0.288)	24 (0.329)
Information-client No. of events: 27 No. of actions: 41	2 (0.074)	8 (0.296)	0	15 (0.556)	7 (0.259)	9 (0.333)
Information-physical No. of events:56 No. of actions: 92	3 (0.054)	20 (0.357)	8 (0.143)	21 (0.375)	17 (0.304)	23 (0.411)
Comment No. of events: 94 No. of actions: 139	12 (0.128)	30 (0.319)	8 (0.085)	25 (0.266)	31 (0.33)	33 (0.351)
Question No. of events: 87 No. of actions: 113	6 (0.069)	40 (0.456)	5 (0.058)	9 (0.103)	29 (0.333)	24 (0.276)
Total action in talk and action events	121	273	82	118	137	153

(The average value of action-type actions per event of talk-type is 0.274.)

3. As might be expected, "Measuring" and "Locating/indicating" actions have high association with "Dimension/location" talk demonstrated by "action values" 0.622 and 0.556 respectively. The other action-types have very low "action values" for "Dimension/location."

4. Not surprisingly, talk-type "Explanation" involves a very high use of "Locating/indicating" actions (0.631) and a moderate use of "Demonstrating function" (0.288), "Examining" (0.288) and "Gesturing" (0.329) actions. The use of "Constructing" actions (0.096) is very low.

5. "Demonstrating function" ("action value" 0.556) actions are more commonly used in "Information-client" talk than in other action-types. Moderate use occurs of "Gesturing" ("action value" 0.333) and "Locating/indicating" ("action value" 0.296). No "Measuring" actions occur with "Information-client" talk.

6. "Information-physical" talk is accompanied by above-average use of action-types "Locating/indicating" (0.357), "Demonstrating function" (0.375), "Examining" (0.304) and "Gesturing" (0.411). Action-type "Constructing" has very low use.

7. The action-types "Gesturing" (0.351), "Examining" (0.33), "Locating/indicating" (0.319) are commonly used in talk-type "Comment".

"Demonstrating function" (0.266) use was less common. "Constructing" actions (0.128) and "measuring" actions (0.085) were rarely used.

8. Talk-type "Question" shows a moderate use of "Locating/indicating" actions (0.456), "Examining" actions (0.333) and "Gesturing" actions (0.276). The other action-types are accompanied by low "action values," indicating very low usage in this talk-type.

The action patterns observed in the data of Table 6.7 prompt the following comments:

1. The results show that the work of "Design proposal" attracts the highest usage of two particular action-types, "Locating/indicating" and "Constructing."

2. "Locating/indicating" actions are important to all the talk-types; they have a wide spectrum of application and a very high usage in the talk-types that could be considered to offer explicit information.

3. "Measuring" has a narrow spectrum of use. The high use of "Measuring" action in talk-type "Dimension/location" was expected, but its lack of use in other talk-types is perplexing. It would reasonably be expected that "Measuring" would feature strongly in "Design proposal" because measuring provides specific detail about size and location. The zero use of "Measuring" in "Information-client" is a reasonable expectation because the physiotherapy and occupational therapy assessment when physical features of the client were examined was excluded from the videotape acts and vignettes analyzed.

4. "Design proposal" and "Dimension/location" talk have high association with action-type groupings, "Constructing" and "Locating/indicating," and "Measuring" and "Locating/indicating," respectively. In both talk-types the pair of action-types with a very high use contrast with the below-average involvement of the remaining action-types. The high "action value" indicates that these action-types appear to have a specific association with these talk-types. For example, "Constructing" actions are clearly associated with "Design proposals," "Measuring" actions are clearly associated with "Dimension/ location," and "Locating/indicating" actions are important for both talk-types. The other action-types do not appear to be as important in communicating "Design proposal" information and "Dimension/ location" information. The dichotomy is not repeated in the other talk-types.

5. The high "action value" for "Locating/indicating" associated with "Explanation" talk and high "action value" for "Demonstrating function" associated with "Information-client" talk suggests that information communicated in these talk-types is reliant on certain types of actions. These talk-types show near-average "action value" for most of the other action-types, hence they are different from the case above for "Design proposal" and "Dimension/location" talk.

6. "Gesturing" has less specificity than the other action-types and enjoys almost universal application across talk-types. Its near-average "action value" in six of seven talk-types suggests that "Gesturing" plays an important role, although at this stage of analysis the role is elusive. It may be used as a social modifier to attract and maintain attention.

The "value" of action to talk

The importance of action to a corresponding piece of talk was determined by considering the understandability of participants' talk with and without observing their actions. In multiple-action events, the influence of all actions on understanding was considered. Each action was not individually considered. Four outcomes of listening were considered:

7. If the talk on the videotape (audio with no images) made no sense and, on replaying the tape, observing the action accompanying the talk enhanced the understanding of the talk, then the action was rated as "gives meaning" to talk.
8. If the talk on the videotape made sense as English expression, but the meaning of the talk was unable to be established without seeing the action, then the action was rated as "identifying the talk." For example, the action could clarify what part of a wheelchair was being talked about or the location of an object.
9. If the talk alone made complete sense and that sense was unchanged after seeing the action, the action was rated as "embellishing the talk."
10. If the action appeared to be unrelated to the talk, it was given a rating "unrelated."

The "Role of action" in the talk of the analyzed videotapes is presented in Table 6.8. The results indicate that in 79% of the events, seeing the action was necessary to understand what was said (in 44% of events the actions "give meaning" to talk and in a further 35% of events seeing action is needed to "identify the talk"). Action embellished an event in 13% of cases. In 8% of events the action appeared to be unrelated to that talk in the event.

Considering Tables 6.7 and 6.8, events coded "Design proposal," "Dimension/location," "Explanation," "Information-client," and "Question" contain large amounts of action accompanying the talk and also manifest a high percentage of occasions where the action is critical to the talk in either giving meaning or identifying something in the talk. These outcomes indicate the extensive use of action with talk by the participants to enhance their communicating. The outcomes also indicate the need for participants to see each other's actions in order to gain the full value of each person's talk.

Table 6.8. Role of action in relation to talk

Talk-type	No. of events	"Role of action"			
		Gives meaning	Identifies talk	Embellishes	Unrelated
Design proposal	138	80 (58%)	36 (26%)	16 (12%)	6 (4%)
Dimension/location	45	24 (53%)	9 (20%)	9 (20%)	3 (7%)
Explanation	73	33 (45%)	20 (27%)	11 (15%)	9 (12%)
Information-client	27	14 (54%)	11 (38%)	2 (8%)	0
Information-physical	56	19 (33%)	22 (40%)	9 (16%)	6 (11%)
Comment	95	30 (32%)	39 (40%)	14 (15%)	12 (13%)
Question	87	27 (31%)	45 (52%)	9 (10%)	6 (7%)
Total	521	227 (44%)	182 (35%)	70 (13%)	42 (8%)

Pragmatics and action

It was noticed that participants' speech varied in complexity and consideration was given to investigating whether certain talk-types contained more specific speech than others. For example, "Design proposal" and "Dimension/location" talk should provide detail about sizes, locations, materials, and be highly specific. "Comment" and "Question" talk, for example, would not be expected to contain design-relevant detail; hence, it should be low in specificity. Analyzing talk for the presence of nouns, pronouns, and adverbs was appropriate as a simple test of an utterance's descriptive specificity. Table 6.9 records the use of nouns (naming words providing specificity), and pronouns and adverbs (words reducing the preciseness of talk), in these four talk-types sampled across the videotapes. Action-type and the "Role of action" of each sample was also recorded.

Table 6.9 provides evidence that: (1) High specificity talk-types "Design proposal" and "Dimension/location" have a high incidence of Talk & Action. Lower specificity talk-types "Comment" and "Question" have a mix of talk, and Talk & Action. (2) "Design proposal" talk-type averages more words per event than the other talk-types. However, in each talk-type the structure of sentences contains large numbers of pronouns and adverbs compared to nouns. (3) "Locating/indicating" actions have high incidence across all four talk-types, and a very high percentage of events in each talk-type gain lucidity only by viewing the action accompanying talk. These results indicate that

Table 6.9. Summary of talk-type, role of action, action-types, and vocabulary in a sample of design events

Talk-type	Role of action	Major action-type	Numbers of words in events	Statistics for noun/pronoun/adverb in events
Design proposal 3 talk only 32 talk & action	3 Embellish 7 Identify 22 Gives meaning	24 Locating/ indicating 15 Constructing 15 Gesturing	av = 30.31 s = 17.2 n = 35	Σ = 118 / 78 / 48 n = 31, 28, 24 av = 3.81, 2.79, 2.00 s = 2.14, 1.87, 1.02
Dimension/location 27 talk & action	4 Embellish 7 Identify 13 Gives meaning 3 Unrelated	17 Measuring 13 Locating/ indicating 6 Examining	av = 16.96 s = 14.86 n = 26	Σ = 65 / 34 / 35 n = 23, 15, 18 av = 2.83, 2.27, 1.94 s = 2.57, 1.22, 1.39
Comment 12 talk 24 talk & action	2 Embellish 10 Identify 11 Gives meaning 1 Unrelated	12 Locating/ indicating 9 Examining 7 Constructing 7 Gesturing	av = 14.31 s = 7.67 n = 36	Σ = 47 / 48 / 38 n = 24, 31, 24 av = 1.96, 1.55, 1.58 s = 1.08, 0.68, 0.78
Question 15 talk 22 talk & action	10 Identify 11 Gives meaning 1 Unrelated	10 Locating/ indicating 11 Examining 6 Gesturing	av = 11.24 s = 6.25 n = 37	Σ = 47 / 43 / 38 n = 25, 25, 22 av = 1.88, 1.79, 1.73 s = 1.13, 0.72, 1.03

Note
av: the average in the applicable events.
Σ: the sum of the relevant item in the applicable events.
n: the number of applicable events.
s: the standard deviation around the average.

team members speak with a simple vocabulary and add a visual channel to their communication to illustrate certain elements of the oral delivery.

Discussion

The results of this research indicate that design talk does not occur in isolation from artefacts and actions involving artefacts. The major role of action for the participants of the seating clinics was to make practical the ideas they wished to express. In some cases action was performed to illustrate to listeners things that were difficult to express in words, thereby enhancing the understanding of talk and making competent a communication. In other cases action was performed more as a private affair to help the actor clarify things in his or her mind by providing the opportunity to test ideas, check relationships, and visualize cognitively difficult concepts. Action performed by a speaker frequently induced the involvement of the other participants. Participants were observed to act in unison, to finish another's action, and to complement another's action, which is similar to verbal interaction that occurs between two speakers. It was observed that an idea proposed by one person underwent enhancement during its journey through a to-and-fro exchange between two, sometimes three, participants. This idea enhancement is reminiscent of a tag team where one participant takes the idea, works with it, then passes it on, or has it taken up by the next player who works some more before releasing the modified idea.

Artefacts

Approximately 83% of actions in the Talk & Action events involved an artefact. The artefacts employed by clinic participants to assist their actions were frequently the elements associated with the current discussion – for example, an armrest on a wheelchair, seating hardware the speaker manipulated with respect to the client's body, or part of the client's body. Artefacts could also be things at hand such as a piece of polyurethane foam or an off-cut of plywood. Occasionally, it was observed in videotapes that a participant would retrieve an artefact from elsewhere and bring it into the discussion. Such artefacts probably better suited the concept to be communicated or provided a better opportunity to develop ideas through constructing an impromptu prototype or more appropriate test of an idea. Figure 6.4 shows an engineer and technician discussing options for attaching a chest restraint strap to the backrest of a seating system. A length of seatbelt webbing wrapped around a pen is being used by the technician to mock up a solution and test various locations and means of anchoring the strap. Simple artefacts and their assembly into impromptu prototypes can be as meaningful in context as dedicated, sophisticated hardware components.

Team members using artefacts, individually and in groups of two or more, gathered around the wheelchair discussing and trying out ideas, helped to build a common (shared) reference and mutual understanding about design possibilities, design and manufacturing decisions (specifications), and achievable outcomes. Artefacts served to focus the attention of the clinic participants, stimulated question–answer dialogue, and drew out participants'

Figure 6.4 Rehabilitation engineer and technician discussing options for attaching a shoulder strap. (The technician has made a loop in the strap and placed his pen through the loop to represent a stainless steel rod.).

experiential knowledge and considered opinion. Observation of many designing episodes where it was evident that there was a close association of artefacts at hand with an idea or sequence of ideas proposed by team members, provided evidence of the roles of artefacts in aiding the design process. These were: (1) to initiate discussion and discovery; (2) to stimulate the generation of new ideas; (3) to verify information and understanding, and (4) to develop and propose ideas. Combining artefact and action enabled team members to impromptu prototype. The rapid application of artefacts such as things at hand provided immediate feedback to team members. Demonstrating the relationships of components, exploring the viability of design options, the preliminary planning of manufacture, illustrating to a client and care-givers what a final product would look like and its working, were the benefits of impromptu prototyping. The high usage of artefacts in rehabilitation engineering work is an indication that artefacts add value, probably because they assist participants to increase the complexity of data communication through creating visual reference points to underpin oral descriptions. Harrison and Minneman (1996) made similar observations and stated that objects were often introduced into conversations for the express purpose of illustrating a particular quality that could not be addressed directly solely by talk or talk plus illustrating with sketches.

A little recognized value of artefacts and the artefacting environment is the reminder value of an important artefact. Whittaker and O'Conaill (1997) called this "context-holding value." In some videotapes of REC practice there was regular revisiting of unsolved problems. It was possible that the sight of the client or the continued presence of an item of hardware had "context-holding value" for the team, acting as conscious reminders of difficult things to be done.

An artefact as an inanimate physical object may have no inherent meaning in the context of a design scenario. In these cases it is only when a person does something with the artefact that it acquires meaning. In many cases this meaning will be enhanced by what the user of the artefact says as it is put to use.

Triggers to action

Harrison and Minneman (1996) suggested the triggers to interaction with objects they observed in a small design team: to seek information, to control the dynamics of a conversation, to change topics, to confirm or recalibrate imaginary objects. The main triggers to action with objects operating for people working at REC that were observed in the videotapes were:

1. as props to illustrate or demonstrate something that seemed too complex to deliver solely by talk;
2. where an idea was being tested; and
3. as a means to plan a strategy or a way forward through a problem.

Words, deictics, action

Pronouns like "it, this, that" and adverbs like "here, there, up" were commonly used words in the talk. The majority of these events contained a high incidence of "locating/indicating" actions, providing a reference for the talk's focus. Harrison and Minneman (1996) also noticed the designer's frequent use of the vague terms "here," "this," and "there," and the pointing, holding and making of shapes with fingers and hands that occurred in concert with talk. Tatar (1991) commented that: "the success of a deictic reference depends on the shared knowledge about the position of the object."

In this study it was evident that team members preferred to incorporate imprecise vocabulary in their speech rather than select words with precise meaning (nouns). Our thesis is that team members engage in artefacting because the combination of talk, action, and artefacts provides more detailed (highly specific) design information to other participants than would be contained in a purely oral presentation. Artefacting reduces the complexity and sophistication of oral discourse that would be needed to impart the same design information. The talk-types that produce highly specific information – namely, "Design proposals" and "Dimension/location" – occur in events that are predominantly Talk & Action events. Team members utilize all action-types as they talk; however, Table 6.9 provides evidence that certain groups – for example, "Constructing", "Locating/indicating", "Examining" – accompany highly specific talk. Interestingly, "Gesturing" is widely practised across all talk-types. Gesturing may be used as a social modifier and to attract and maintain attention. It is obvious that, for successful communication to occur when pronoun- and adverb-intensive talk is used in design conversations, speakers' accompanying actions need to be seen clearly by the listeners.

A notion of observing and experiencing artefact and action

For seating clinic participants artefacts performed the "compounding roles" suggested by Dix (1994) – that is, rather than have an isolated impact on conversation, artefacts can have an accumulating effect, adding layers of meaning. Artefacts were used by individuals, shared among team members, and the individual and shared use was available to be observed by other participants, both engaged and non-engaged in that work at the moment. In use artefacts provided feedback to users, and provided "feedthrough" for observers as they viewed the effects of action by the users. Participants were able to achieve mutual understanding from their various artefact interactions, which Dix called "soft artefact." Participants were able to communicate through interacting with artefacts and by referring to artefacts – hence, artefacts engendered a spirit of cooperation between participants.

The resolving of "how?, what?, where?, why?" issues in designing customized seating is assisted by participants of the design process observing artefacts (objects), observing actions, observing actions with artefacts, and experiencing actions on artefacts. Participants are usually performing actions with artefacts or watching another person or other people performing these actions. Figure 6.5 illustrates a model of a talk and action event that can produce an outcome of the "how?, what?, where?, why?" enquiry such as information/knowledge, answers/understanding, ideas, designs. The outcome is derived from observing or experiencing an artefact or an action involving an artefact, and is promulgated by talk or talk and action. The artefact and an action are both the tools for achieving the outcome and aiding communication of the outcome to other participants.

A Model of Talk, Artefacts, Action in Time

The model illustrated in Figure 6.5 was developed to relate the role of talk, artefacts, and actions to generate and communicate an outcome through the observing and experiencing of the artefact and associated actions. Essentially, this model works at an event level. Usually a design evolves over many events

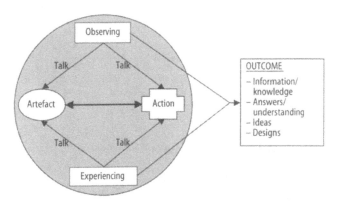

Figure 6.5 A model of the role of talk, artefacts, and actions to generate and communicate an outcome through the observing and experiencing of the artefact and associated actions.

in which participants observe artefacts under action. They talk and develop their ideas, which leads to further talk, action, and artefact generation. Over time, experience and understanding increase as the "observing, experiencing, talking, acting with artefacts" cycle continues. When sufficient experience and understanding have been gained, an acceptable design concept is struck and designing moves on to a new issue. Figure 6.6 illustrates the evolution of this model through experience and time towards design.

Similarity exists between this model and the concepts put forward by Bailetti and Litva (1995) concerning the transformation of customer (client) requirements from informal constraints expressed in natural language into formal statements expressed in design languages. In this model the talk, artefacts and action aid the transformation of informal statements into formal statements. The end of designing in this model is similar to Bailetti and Litva's design process ending, which occurred when the designer produced a formal system model with all the detailed information to realize the required product.

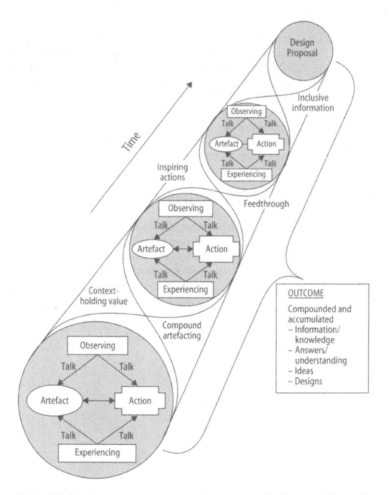

Figure 6.6 Model of the observing, experiencing, talking, acting cycle, through which participants gain experience and understanding to generate a design.

Conclusion

The results of this research provide evidence of the ability of actions to enhance the communication value of talk in designing. Participants of a seating clinic incorporate artefacts into their conversations and act on the artefacts as part of the idea generation process, and apply this action to aid the communication process. Action with artefacts can be inspiring to designers by providing opportunities for serendipitous play as well as purposeful investigative activity within the design space. Speakers place reliance on actions to support and make lucid their talk. The over-arching significance of artefacts is that they aid communication between participants by providing props for a visual channel that supports oral communication. Cooperation based on sharing artefacts is a major strength of face-to-face interaction. Participants can experience artefacts and observe others using artefacts. A simple model is proposed depicting the interaction with artefacts that enhance talk and promulgate further talk based on experience of the artefact or observation of it. The expansion of this simple model with a concatenation of multiple, sequential observations and experiences producing an outcome – an accumulation of information, knowledge, answers, understanding, ideas, and potentially a design – is also proposed.

Notes

1. US Public Law 99–506 defines rehabilitation engineering as the systematic application of technologies, engineering methodologies, or scientific principles to meet the needs of and address the barriers confronted by individuals with handicaps in areas which include education, rehabilitation, employment, transportation, independent living and recreation.
2. Talk transcription was restricted to oral output that conveyed information about the client, equipment, etc. Talking of a social nature, humour, etc. was not transcribed and was not included in analysis.
3. Action is defined as purposeful hand and body movements performed with or without artefacts.
4. An event is defined as the occurrence of action performed in silence, or talk, or talk combined with action by a participant. An event continued for the time a participant was speaking and/or acting.

References

Bailetti, AJ and PF Litva 1995. Integrating customer requirements into product designs. Journal of Product Innovation and Management 12:3–15.

Dix, A 1994. Computer supported cooperative work: A framework. In Design issues in CSCW, edited by D Rosenberg and C Hutchinson. New York: Springer.

Furlong, TJ 1997. Virtual reality sculpture using free-form surface deformation. In Proceedings DETC '97. Paper #DFM4511, ASME Design engineering technical conferences, Sacramento, CA (Sept 14–17 1997).

Goodwin, C and MH Goodwin 1996. Seeing as a situated activity: Formulating planes. In Cognition and communication at work, edited by Y Engeström and D Middleton. Cambridge: Cambridge University Press, pp 61–95.

Harrison, S and S Minneman 1996. A bike in hand: A study of 3-D objects in design. In Analysing design activity, edited by N Cross, H Christiaans, and K Dorst. London: John Wiley & Sons, pp 417–436.

Heath, C 1997. The analysis of activities in face to face interaction using video. In Qualitative research: Theory, method and practice, edited by D Silverman. London: Sage Publications, pp 183–200.

Horton, GI 1997. Prototyping and mechanical engineering. PhD Thesis, University of Queensland.

Kleifgen, JA and P Frenz-Belkin 1997. Assembling knowledge. Research on Language and Social Interaction 30(2):157–192.

Krovi, VG, G Ananthasuresh, V Kumar, and J Vezien 1997. Design and prototyping of rehabilitation aids. In Proceedings DETC '97 paper #DETC97/DFM-4361, ASME Design engineering technical conferences, Sacramento, CA (September 14–17, 1997).

Logan GD 1999. An investigation of talk, action and use of artefacts by a cross-discipline rehabilitation engineering team. PhD Thesis, University of Queensland, 1999.

Luff, P, C Heath, and D Greatbach 1994. Work, interaction and technology: The naturalistic analysis of human conduct and requirements analysis. In Requirements engineering, edited by M Jirotka and JA Goguen. London: Academic Press, pp 259–288.

Peräkylä, A 1997. Reliability and validity in research based on tapes and transcripts. In Qualitative research, edited by D Silverman. London: Sage Publications, pp 201–220.

Radcliffe, DF 1996. Concurrency of actions, ideas and knowledge displays within a design team. In Analysing design activity, edited by N Cross, H Christiaans, and K Dorst. London: John Wiley & Sons, pp 343–364.

Radcliffe DF and P Harrison 1994. Transforming design practice in a small manufacturing enterprise. In Design theory and methodology, edited by TK Hight and F Mistree. ASME Vol. DE-68, pp 91–98.

Radcliffe, DF and P Slattery 1993. Video as a change agent in a cross-discipline design team. In Proceedings: International conference on engineering design (August 17–19, 1993). The Hague: ICED '93, pp 319–326.

Tang, JC 1991. Findings from observational studies of collaborative work. International Journal of Man–Machine Studies 34:143–160.

Tartar, D 1989. Using video-based observations to shape the design of a new technology. SIGCHI Bulletin 21 (2): 108–11.

Tatar, DG, G Foster, and DG Bobrow 1991. Design for conversation: Lessons from Cognoter. International Journal of Man–Machine Studies 34:185–209.

Whittaker, S and B O'Conaill 1997. The role of vision in face-to-face and mediated communication. In Video-mediated communication, edited by KE Finn, AJ Sellen, and SB Wilbur. Mahwah, New Jersey: Lawrence Erlbaum Associates, pp 23–49.

PART III

Beyond Disciplinary Perspectives

8

The Thoughtful Mark Maker – Representational Design Skills in the Post-information Age

Martin Woolley

Skills – Past, Present And Future

Skill, technology, and design

The impact of computing confronts all professions, profoundly affecting the way that they function and challenging occupational relevance and survival. A common misperception is that emerging technologies automatically replace traditional skills, making many specialist skills redundant, effectively deskilling expertise. This chapter explores deskilling in the context of design representation and discusses the potential of new technologies as both a generator and absorber requiring new techniques, approaches, and thinking with which to create a new skills-base.

The application and definition of traditional skills are discussed, with a comparison of the differences that have taken place in the development of design practice between knowledge-based skills and the separation from implementation. This is followed by an analysis of the representational skill-base currently in use in design practice, together with comment on the effects of technological change. Lastly, strategies to develop appropriate representational skills for the future are proposed.

Skills through history

E.P. Thomson describes and evokes an era when the hand drove production and an individual's skills were a complex continuum of manual dexterity, a keen eye, knowledge of the task /role of the object in use, and often direct knowledge of their "clients":

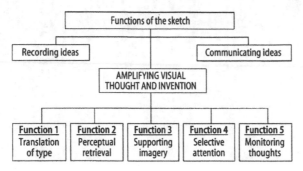

Figure 7.1 Some functions of the sketch.

providing memory search and retrieval cues that improve the availability of tacit visual knowledge for invention.

Hypothesis 3 (support for superimposed mental imagery)
Sketches are percept-memory hybrids. The incomplete, physical attributes of sketches act as a stimulus for percepts that invite completion from memory and imagination. Such properties encourage the generation of transient mental images that, after a matching process, are transformed and spatially superimposed on an internal representation of the sketch.

Hypothesis 4 (selective attention to visual components)
Sketches amplify inventive thought by isolating and representing separately those attributes of visual experience that are of special relevance to a particular task. This amplification assists the user to attend selectively to a limited part of the task, freeing otherwise shared components of cognitive capacity and reducing the complexity of preparatory visual processing.

Hypothesis 5 (conscious monitoring of visual thought)
Sequences of sketching acts support the conscious awareness of one's own cognition. Enhanced awareness assists creativity, providing voluntary control over highly practised mental processes that can otherwise become stereotyped. Unforeseen percepts from untidy or accidental stimuli can elicit unconscious processes which break the mould of habitual thought, while a temporal record of recent ideas makes it easier to change one's mind at appropriate stages.

While it was clear that these cognitive functions were interdependent, I was then embarrassed by the difficulty in finding a single unifying function. Of course, our brains have many modules for independent tasks that are often performed in parallel. So it is possible that the search for a single non-trivial unifying function is vain. Minsky (1985) has suggested that the search for any unified theories of mind is "physics envy." However, physics envy or not, I think I can now see ways by which several of these functions are related in interesting ways. Unfortunately, since the thesis was written, I have discovered two new functions of sketches, so the baffling complexity of our brains is still a major problem. Here I concentrate on showing a connection between the first three functions (Hypotheses 1–3 above).

These hypothetical functions were intended to be compatible with the Baddeley and Hitch model of working memory (Baddeley 1986; Logie 1995).

All the proposed functions were ways of supporting components of working memory and there are good grounds for supposing that working memory is a resource which places limits on the brain's capacity for design thinking (Logie and Gilhooly 1998). It is therefore an obvious candidate for sketch support. Unfortunately, "working memory" – the temporary storage and processing resources used by the designing brain for manipulating internal representations of imagined objects – is itself composed of multiple modules. More is known about separate modules for processing language, visuo-spatial information and for directing attention than is known about how these components interact. Although linguistic and visual memory systems are often treated independently, in design thinking they are interdependent and the brain must possess translation mechanisms between the two that perhaps need cultural support (Paivio 1986). Here I argue that sketch function two (Support for Perceptual Memory Retrieval) and sketch function three (Support for Superimposed Imagery) are both slave functions of function one (Translation of Representational Type). To explain this I will use a metaphor based on chemistry – "cognitive translation catalysis."

The need for such "catalysis" is due to maladaptations in the brain that we have inherited from our Ice Age ancestors.

The Brains We Inherit from Palaeolithic Humans

Why is it that, despite enormous recognition memories, we find it relatively difficult to visualize and mentally manipulate absent objects or events? Why is the limited capacity of our working memory so poorly adapted to a lifestyle and culture that places heavy demands on foresight and planning? In order to design better sketching systems we need to understand not only how our brains work, but also *why* they work the way they do. Some of the answers may be found in evolutionary psychology.

"Is it not reasonable to anticipate that our understanding of the human mind would be aided greatly by knowing the purpose for which it was designed," remarked the biologist George Williams in an influential book on evolution (Williams 1966). By "designed" Williams knew that every biologist would understand that he was referring to the adaptive advantages produced by Darwinian natural selection. Our brains evolved the way they have by the accumulation, by survival and successful reproduction, of small, inherited variations in our nervous systems. In our evolutionary past every genetic part of our brains must have improved the competitive survival rate and fecundity of those individuals who possessed them. Evolution by natural selection is not really design at all, although it looks like design. For such evolution to occur at least two conditions are necessary. The first is long periods of time. The random mutations (copying errors) and recombination of our genes that are necessary to accumulate a significant number of advantageous inherited changes to our nervous systems need thousands of generations of parents and offspring. Evolution usually occurs very gradually in tiny successive changes since large, single-step changes to our genetic make-up are nearly always disadvantageous. The second condition is consistent selective pressure. There must be something about our lifestyle and environment that causes certain genetic variations in our brains to survive and be reproduced more than other variations. Changes to our brains that are caused by learning from the environment or transmitted by culture are not inherited and cannot cause our brains to evolve. But the selective pressure must last long enough to allow enough genetic variation to occur. If the environment and lifestyle change too quickly, then evolution cannot keep pace with the selective pressure, and an organism's anatomy and instinctive behaviour can be maladapted to its present condition. The genetic make-up of an organism can only be understood by considering its evolutionary past. Biologists refer to this as the evolutionary "time-lag" (Dawkins 1982). Because of this time-lag, there may be parts of a species' anatomy and instinctive behaviour that are no longer adaptive. Their existence may even be puzzling to biologists if they are ignorant about the exact conditions surrounding an organism's evolutionary past. Many species survive with a certain amount of such "time-lag." This is normal, but with our species the time-lag is dramatic. The 8000–10,000 years since our ancestors emerged from the last Ice Age, planted the first crops and invented writing, is less than one five-hundredth of the time that it took our brains to evolve from the ancestors we share with our nearest relative, the chimpanzee. The post-Ice Age invention of agriculture, cities, and writing has occurred in too short a time to allow a significant evolution of the brain. And the Industrial Revolution with its concomitant separation of design from manufacture is less than the flicker of an eyelid in biological time. Yet the last hundred years have witnessed a hundredfold growth of scientific knowledge (Ziman 1976, p. 56) and a demographic explosion that have caused massive changes to our environment and lifestyle. There is no known mechanism by which our genes can have adapted to cars, city planning or computers. Thus, our brains evolved for solving problems that were quite different from those that our culture now imposes on it. Although there probably have been small genetic changes to our nervous systems in the last 10,000 years, archaeologists consider our upper palaeolithic ancestors, who lived in Europe and Asia during the Würm glaciation between 35,000 to 10,000 years ago, to be anatomically modern humans. They possessed a brain that was the same size and shape as ours. It is to their lifestyle rather than ours that our brains are adapted. Consider what can happen to our brain–body systems when we are driving in heavy traffic. Responding to stress, our brains cause adrenalin or noradrenalin, the hormones of aggression and fear, to be circulated in our bodies. This process in turn causes microglobules of fat to be squirted into our blood. This adaptation increased the survival chances of our ancestors, providing much needed energy to our muscles for attacking a mammoth or fleeing a lion. However, for the modern motorist sitting in a car, the fat piles up in the heart and arteries, causing one of the fastest growing causes of mortality in Western culture.

A time-lag, less obvious perhaps and poorly understood, also exists in the neural resources we are endowed with for thinking about and imagining future objects and events. It took $2\frac{1}{2}$ million years for the brains of the first tool-making species of the genus Homo (*Homo habilis*) to increase in size from an average 600 cu. cm. to the 1350 cu. cm. we possess today (Lewin 1993). In most groups of related mammal species, the ratio of brain size to body surface is approximately constant. If this ratio is used as an "index of encephalization" (Bauchet and Stephan 1969), then the index for most primates (including our nearest relative) is 11.3 times that of an insectivore. However, the index for modern humans breaks the primate rule and is $2\frac{1}{2}$ times greater. The greatly expanded neocortex, with its much larger prefrontal lobes that we possess compared to other primates, was an evolutionary adaptation to a nomadic hunter-gathering lifestyle (Leakey 1994; Bradshaw 1997). These changes favoured in turn the natural selection of visually controlled manual dexterity, high capacity long-term memory for places and objects, and the ability to react quickly and flexibly to unexpected dangers and opportunities. The brain evolved complex, innate mechanisms for using stored knowledge to process percepts and plan motor actions. The neural regions associated with foresight and planning became greatly expanded. However, survival depended more on the brain's ability to make fast, flexible responses to unexpected opportunities and dangers than on its capacity to plan for distant futures. An earlier species of our genus, *Homo erectus*, learned to domesticate fire 400,000 years ago and the Levallois technique for working sharp symmetrical flint tools from many flakes from the same stone is 250,000 years old (De Lumley 1998). By 35,000 BP the modern *Homo sapiens* of Ice-Age Europe made sophisticated, decorated tools from stone, bone and clay, cooked food, ceremonially buried their dead and made extraordinary paintings on the walls of caves. It must be remembered throughout the period 35,000 to 10,000 BP our "anatomically modern" subspecies, *Homo sapiens*, was a rare animal. Archaeologists estimate that the total population in France during the Magdalenian era (19,000–11,000 BP) was not more than 50,000 and the total population in the world perhaps less than one million (De Lumley 1998). A huge biomass of fauna and flora was available for food. Humans lived by gathering fruit and edible roots, and hunting reindeer, horses, wild oxen, bears and ibex. Many other species of game existed, now extinct in Europe, such as lions (which were bigger than today's animals) and bison, or are extinct in the world, such as the woolly rhinoceros, mammoths and the humpbacked megaloceros. Climate was an important factor in the evolution of our brains. The weather was often so cold that a glacier 1.5 kilometres in diameter covered the whole of northern Europe and caused sea levels to fall 150 metres. Clearly, very large memories and great inventive tool-making skills were necessary to survive. Ample evidence of creative intelligence is provided by the artefacts that have survived. The invention of the technique for calcining flint, making it sharper and easier to work, the invention of the spear-thrower and the barbed harpoon all come from this period. So also does the invention of sewing (c. 20,000 BP) when eyed needles as fine as some modern ones were ground from ivory using only a flint awl and a specially designed grooved stone (Musèe de l'Homme, Les Eyzies, France). The "Venus de Brassempouy" is four times older than the invention of writing (Lewin 1993). It is a tiny but perfect ivory carving of a young woman wearing a coiffure or head-dress worthy of St Laurent. A palaeolithic sepulchre at Sounguir, Russia, shows the ceremonial burial of a 40-year old man dressed in an elaborate coat with sleeves and trousers with shoes attached. The material has long since disappeared, but the shapes of the clothing can be seen from the beautiful lined decoration of beads, with bracelets and head-band, which remains (De Lumley 1998). This evidence and much more (White 1999) makes it easy to believe that the hunter-gathering brain of the upper palaeolithic was at least as intelligent and creative as ours. Clearly, there existed a sophisticated culture four times older than writing.

Two inherited components of our mental resources are especially relevant to design thinking – the "language instinct" (Pinker 1994) and the ability to make visual images, "the visualizing instinct." In order to design better sketching technology, we need to understand which capacities of our brains are innate and which must be learned or culturally augmented.

The Language Instinct

There is good evidence that we have innate cerebral resources for acquiring and using spoken language both to communicate ideas to others and to represent them to ourselves. Of course, we are not born with knowledge of a particular syntax and vocabulary. However, as Chomsky (1957, 1980) has shown, all languages share a common "deep structure." Pinker (1994) surveys persuasive evidence to show that children's ability to acquire and use language is innate. For example, he quotes a study that shows how children of immigrants can spontaneously improve on a pidgin language, the only language they have been exposed to. Pinker also quotes an astonishing discovery that a new Nicaraguan signing language (ISN) was "created in one leap" by deaf children when "the younger children were exposed to the pidgin signing of the older children." ISN has a consistent grammar that in many ways is superior to the pre-existing official signing language. Further, "ISN has spontaneously standardized itself; all the children sign it the same way."

There are roughly 5000 languages in the world today. By studying the similarities and differences between existing languages it is possible to infer an evolutionary family tree of language origins from earlier, now extinct root languages. It is also possible to infer an evolutionary family tree of our genetic origins by studying the DNA sequences of tissue from different races all over the world. Interestingly, these two trees, derived from completely unrelated sources, show a close match (Cavalli-Sforza 1991). Such a parallelism between genetic and linguistic data is consistent with the evidence for a genetic component to language. The cultural evolution of language tracks our genes.

We are born with a complex of neural circuits in our brains that underlie the ability to acquire and understand spoken languages (Pinker 1994). Thus, Broca's area, in the frontal lobe, is necessary for the grammatical production of speech. Further back in the cortex, Wernick's area is involved in speech understanding. The two areas are connected but other parts of the brain are also necessary. Brain-imaging studies show language activating many circuits of the brain and its neural basis still uncertain (Pinker 1994). Exactly when, in hominid evolution, language first appeared is also uncertain. Moulds of the inside of the *Homo habilis* skull 1470 show clear evidence for the existence of Broca's area 2 million years ago (Leakey and Lewin 1992). However, this area may have changed its function since then. More persuasive evidence for an early origin of language comes from a curvature of the base of the cranium in fossil skulls. This has been shown to correlate in primates with how low the larynx is in the neck. A low larynx is necessary to provide the large pharyngeal space needed for speech. However, a low larynx carries a heavy penalty. It prevents simultaneous breathing and swallowing. Evidence for a fully flexed basicranium (and thus a low larynx) is found between 400,000 and 300,000 years ago in what is called archaic *Homo sapiens* (Laitman 1983). The low larynx must have had strong selective advantages to override its disadvantages. What advantage could this have provided other than that of spoken language? Since speaking leaves no traces, the evidence for prehistoric language must be indirect. However, there is evidence for trading and social communication of a kind that implies the use of language in the upper palae-olithic humans (de Beaune 1995). Most archaeologists agree that, by then, humans had developed spoken language, although how sophisticated it was and how it was used can only be guessed. As Pinker argues persuasively, the language instinct probably evolved slowly and gradually by the Darwinian process of tiny successive steps, from vocal sounds with simple meanings to the conceptual abstraction and the rich, recursive phrase structure of today's languages.[1]

Experts are still debating the adaptive advantages that language conferred on our ancestors. The obvious first function of language is communication. Language must have coevolved with the instinct for social cooperation in groups, providing the advantages of collaborative hunting of large game and the ability to communicate information about food sources for scavenging or gathering. Another theory suggests that language replaced the grooming that is used for social bonding in other apes. According to Dunbar (1996), neocortex size, group size and language coevolved, as language replaced grooming in larger groups as a means of cementing friendships to ensure future cooperation and to learn how other individuals were likely to react to a given behaviour. In support of his theory, Dunbar has shown that there is a linear correlation in primates between cortical ratio (the ratio of the size of the neocortex to the size of the rest of the brain) and group size. The human cortical ratio is 50% larger than the chimpanzee's. Using this correlation he was able to predict that human group sizes should be 150 (on average). His predicted data agree well with what is, in fact, found in modern hunter-gathering societies but is, of course, greatly exceeded in technological cultures. Is this another example of time-lagged brains?

Not every expert agrees that communication was the primary advantage driving language evolution. Jerison (1991) has argued forcefully that the primary reason for language evolution was not communication, but concep-tual thought. Language is needed for sequential reasoning about possible out-comes of a planned action. Jerison argues that inner reflection and the power to elicit private imagery, not outer communication, was the facility that drove language evolution. Communication was a useful consequence of this facil-ity. It seems to me that the debate about which came first, communication or conceptual thought, is vain. Selection would surely have worked simultane-ously on both communicating and reasoning skills, making them evolve in tandem. The key point is that verbal thought can be verbally shared with others, but that when we think with mental imagery there is no comparable innate mechanism for sharing our imagery.

Archaeologists have been divided over whether the increase in the size of the hominid brain was driven by the needs of hunting and tool-making or by the advantages of language (Leakey 1994; Lewin 1993). Recently, it has been suggested these three activities may have shared neural resources and a common origin. Calvin (1993) points out that activities such as the skilful striking of a flint or the accurate throwing of a spear occur too quickly to be controlled by neural feedback and correction. The brain needs sequence buffers in which the complex sequence of muscle commands can be prepared and mentally tested before such actions can be executed smoothly. Calvin sug-gests that neural circuits that evolved for planning skilled movements could have been adapted later by evolution for planning the sequences of phonemes and words necessary for speech. "An elaborated version of such a sequencer may constitute a Darwin machine that spins scenarios, evolves sentences and facilitates insight by off-line simulation" (Calvin 1993). Earlier, Corballis (1991) proposed that one reason why 95% of right-handed persons have both handedness and language controlled by the left side of the cortex is that both language and manual dexterity use shared mechanisms of non-spatial "praxis." Studies of early hominid stone tools show that they were made by people with lateralized brains like ours (Leakey and Lewin 1992). Interestingly, it has been shown recently with brain imaging techniques that Broca's area, long thought to be specialized only for language production, is also active during manual control (Blackmore 1999).

The theme that mental capacity often depends on "borrowing" and adapting neural resources that evolved for a different function is central to the theory of sketch function. In the Baddeley and Hitch model, a specialized component of working memory, the "phonological loop" is used for the temporary storage of words and symbols during language production and comprehension. It is also used for counting, mental arithmetic, and problem solving. Baddeley and his colleagues have distinguished two subcomponents, a passive acoustic store and an articulatory rehearsal process. The traces of remembered words continually fade in the acoustic store and must be revived by the rehearsal process. The capacity of this store is only five to seven items. When we reason to ourselves about categories and concepts we use an "inner voice" and an "inner ear" (Baddeley and Lewis 1981). Such a sound-based memory system may be well adapted to the quick response needs of a hunter-gathering lifestyle: "Is anyone missing?", "Watch out, the bison is going to charge," etc. When fast responses are needed, a low-capacity, fast-fading word store might even help the brain to prepare for the next message. However, such a short-lived memory store is hardly well adapted to the sequential reasoning needs of a reflective, symbol-processing culture. The very fact that linguistic working memory is based on the sounds of words rather than their symbolic meanings is evidence for an evolutionary time-lag. There is no innate working memory store specific to reading and writing. Of course, we can and do supplement such a poor facility for conceptual reasoning with visual working memory (e.g., Hayes 1973). However, as I argue in the next section, visual working memory is also time-lagged and, besides, designers need their visual working memories for purposes other than symbol processing.

The Visualizing Instinct

We have inherited from our primate ancestors one of the most sophisticated visual systems in the animal kingdom. The task of analyzing the confusing light patterns presented to our retinas is parcelled out to about 34 separate specialized regions in the cortex of our brains (Zeki 1993). Many of these contain maps of either the image on the retina or of the visual field. It has long been known that our brains must supplement the information provided by our eyes in order to reconstruct the visual scene of meaningful three-dimensional objects that we experience when we say we "see." As the reindeer runs away, the image on the hunter's retina gets smaller in direct proportion to the distance. But the Stone-Age hunter sees his game is staying the same size and accurately perceives its changing distance and speed. His or her survival depends on this ability. The colour of berries may determine whether they are poisonous or good to eat. The visual system must "deduce" the spectral composition of the lighting and subtract this from the reflected light to allow us to "see" the real colour of the berries.

Consider the palaeolithic hunter aiming his spear at an antelope. The light may be poor and the image on the retina hazy or too brief for detailed processing. The animal will be camouflaged and it may be partly concealed behind a bush. According to a theory I have documented elsewhere (Fish 1996), the visual system handles such incomplete and confused images in two stages. First, it must identify the animal from the bits that are visible by referring to long-term memory (recognition by parts). Second, it must use a stored representation to generate internally a spatial depiction of the animal. The mental image is then rotated and scaled to match the image from the retina so that it can be superimposed in spatial register with its neural representation to create a hybrid image partly from memory and partly from the eyes. Thus the hunter instinctively uses unconscious knowledge to guide the spear. I speculate that there must be twinned images from memory and the eyes which may be interleaved with spatially corresponding parts, much like the twinned images from the two eyes that are interleaved and fused, if they are compatible, for binocular vision. There must then be a spatial comparison process and, as with retinal rivalry, a gate that allows a hybrid image to be constructed if the two images match, but which allows only the stronger of the two images to become conscious if the match fails. This account of percepts as hybrid images is controversial, but the evidence that memory can influence the perception of impoverished stimuli is not. In an early experiment that fits the gating mechanism proposed above, Bruner and Postman (1949) asked subjects to identify ordinary but hand-painted playing cards exposed for a fraction of a second. Unknown to the subjects a few of the cards had the incorrect colour for their suit, black diamonds or hearts, red spades or clubs. The authors recorded four types of response to the incongruous cards. The first they call a "dominance" reaction in which the subject reports with assurance that a red six of spades is either a normal (black) six of spades or is the six of hearts, depending on whether the colour or shape is dominant. Another type of response they call "disruption" in which the subject fails to resolve the stimulus and does not know what it is (although at the same exposure time he *can* identify normal cards). In a third type of response the incongruity is "recognized." However, in a fourth type of response, there is a "compromise." For example, a subject may report that (a) the red six of spades is either the purple six of spades or the purple six of hearts; (b) the black four of hearts is reported as a "greyish" four of spades or (c) the red six of clubs is seen as "the six of clubs illuminated by red light". Analogous experiments have demonstrated the influence of memory on the perception of colours (e.g., Bruner et al. 1951). There are other examples in the literature where an indeterminate stimulus is completed with information from memory (e.g., Segal 1972; Segal and Nathan 1964; Farah 1985).

Experimental evidence shows that the ability to generate images from memory of absent objects shares many of the properties and format of perception (Finke 1980, 1985; Kosslyn 1980, 1994). It is clear that the resources that our brains provide us with to imagine non-existent objects are "borrowed" from and shared with the neural machinery that evolved for perception (Kosslyn and Sussman 1994). It has frequently been observed that damage to a specialized component of the visual cortex for colour or shape, for example, produces a corresponding inability in the power to imagine (Farah 1985, 1988). Even more striking is the evidence from brain-imaging studies. Depending on the nature of the imagery task, it has been shown that carefully matched imagery and perception tasks cause differential regional blood flow in the same parts of the cortex (Kosslyn et al. 1997). A spatial imagery task activates the same region in the parietal cortex as a corresponding spatial perception task. Interestingly, it is not only visual cortex that is shared with imagery. When a subject mentally rotates one of two shapes into congruence with the other in order to make a comparison, the same region of the motor planning cortex is activated as when we move our hands. When a joystick is rotated manually simultaneously with the mental task, it speeds up the rate of mental shape rotation if it is in the same direction and slows it down if it is in the contra-direction (Wexler et al. 1998). Kosslyn and his colleagues have found that, with detailed imagery tasks, even the primary visual cortex, the very earliest part of the visual pathway to map the information from the retina, is active (Kosslyn et al. 1999).

Thus the capacity to generate, inspect and manipulate mental images of imaginary objects is an incomplete time-lagged adaptation of our visual systems and our movement planning systems. Because of the way our brains work, untidy or incomplete visual stimuli can both elicit and support mental imagery. Without such support, visual imagery is hazy and fades quickly. It is as if mental images were stored with perceptual parts "looking for" incoming images to repair. As with words in the phonological loop, images in visuo-spatial working memory need constant refreshment (Phillips and Christie 1977). As Kosslyn (1994) has suggested, the fast-fade characteristics of the shared "visual buffer" may be an adaptation to the fact that our eyes move in short "saccadic" jumps every fifth of a second. Passing images in the buffer must be quickly erasable to make way for new ones. However, this property is not well adapted to slow reflection about imagined objects. It needs cultural support.

Between 35,000 and 10,000 BP our hunter-gathering ancestors produced, in uninhabited inaccessible caves, beautifully observed paintings and sculptures of animals that included perspective and the representation of movement. The shaman theory of Clottes and Lewis-Williams (1996), although not accepted by all experts, is, to my mind, the only convincing explanation of the properties and circumstances of upper palaeolithic cave paintings. Shamans are spiritual leaders and healers who believe they can communicate with, and even unite with, the spirits of certain animals and access their spiritual power. Shamans have been known since Marco Polo and are found wherever there are, or have been, hunter-gathering people, in North and South America, in Siberia and South Africa where David Lewis-Williams has studied the cave art of the San people of the Kalahari over many years. To the San shamans, the cave walls were never regarded as a passive medium, but were thought to be a gateway to the underworld where faint images of the animal spirits could already be seen. Trained to put themselves into a hallucinatory trance state, shamans believe that the animal spirits enter their bodies and control it as they engrave or paint their images. The elderly daughter of a San shaman told Lewis-Williams that her father, by placing his hands on the cave wall next to the anima image, could extract power from the spirit or heal a sickness. Stencilled images of hands are often found associated with both modern and prehistoric hunter-gathering cave paintings. In the prehistoric "Grottes de Chargas" in the Pyrenees there are 217 such hands.

Experts agree that, whatever their function, these paintings incorporated and made use of accidental images in the bulges, fissures and scratches on the cave walls (Lorblanchet 1999). As Lorblanchet says "it is as if the caves were already inhabited by the spirits of the animals they knew." Figure 7.2 shows a running ibex with legs from natural grooves in the wall from the "Grotte de Cougnac," France.

In the "Grotte de Peche Merle" there is a beautiful horse with its head in a natural head-shaped rock formation. The shape of many animals follows the line of the relief structure of the wall. The "Grotte des Combarelles" (Les Eyzies, France) is a long tunnel into a hill decorated along its length with engraved animals inspired by natural forms on the rock walls. At the far end of the tunnel (and therefore nearest to the underworld) there is a single stencil of a child's hand. The skill with which the natural rock structure is used is seen vividly when the guide switches off the electric lights in the tunnel and uses his cigarette lighter to imitate a prehistoric oil lamp. By its flickering light the shadows cast by the natural relief make the animals spring to life and seem to move.

But there are also places on the cave walls where our prehistoric ancestors' animal images are mixed up with random finger marks in what was once soft clay (Figure 7.3). Thus palaeolithic artists had anticipated by 30,000 years Leonardo da Vinci's theory of the deliberately untidy sketch: "For confused things rouse the mind to new inventions" (Gombrich 1966; McMahon 1956).

This history is relevant to design representation because our culture has provided us with many techniques for adapting our brains to new tasks in ways that we do not yet understand. Sketching is one such technique. In order to manipulate representations of imagined objects in working memory our brains appear to "borrow" temporary storage resources that evolved for recognizing and acting on real objects. Thus the fast-fade characteristics of short-term visual memory (Phillips and Christie 1977) is well adapted to the need to process without interference a stream of rapidly changing images from the

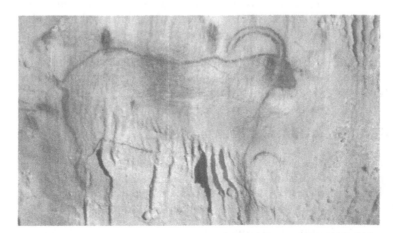

Figure 7.2 Ibex from the Grotte de Cougnac, France, 22,000 BP (from Clottes and Lewis-Williams 1996).

Figure 7.3 Grotte des Trois Frères, France. Detail from a wall of superimposed drawings and finger marks (from Clottes and Lewis-Williams 1996).

retinas of our moving eyes. The need to convert remembered form to the same structure and format as a sensed form in order to make a comparison is, I suggest, a key to an understanding of visual thought. When design becomes separated from manufacture we are forced to invent substitute stimuli that allow us to adapt visual resources to tasks for which they have not evolved.

However, just as the resources that are available to manipulate imagined objects without visual support are inadequate, so too our brains are poorly adapted to manipulate covertly verbally represented abstractions without external support. The capacity limit of four to seven "chunks" for verbal working memory is not enough to support complex sequential reasoning. Our inherited brains are well adapted by evolution to use stored knowledge to solve problems presented by immediately present sensory information or which demand immediate motor responses. But to manipulate stored information within our brains to solve problems about imagined future objects, our brains are forced to "borrow" resources that evolved for perceiving, acting and communicating. Hence the need for behavioural substitutes – representational systems that can be perceived and physically manipulated such as writing and drawing – and hence also the sense of "conversing with oneself" using notes or sketches.

Cultural Evolution: Sketches as Memes

I have tried to show that the two principal innate neural media that evolution has provided for thinking – language and mental imagery – are maladapted for design representation in a science-based culture. Undoubtedly, our Ice-Age ancestors needed to visualize immediately immanent events, how to strike a flint-tool, where to find new roots and berries, what to expect when hunting a mammoth. No identifiable design representations have been found among thousands of palaeolithic images. When design is concurrent with making, the object itself supports the image. The prehistoric paintings and sculpture show that by 35,000 BP there was already a capacity to imagine important and familiar objects, but that this capacity was still tied to the visual system in the brain and needed visual support.

However, there is a major compensation for these disadvantages. Unlike that of any other primate, the human brain increases more than fourfold after birth (Bradshaw 1997). Despite very similar gestation periods, the brain of our nearest relative, the chimpanzee, increases only 60% after birth. Thus the human brain's capacity to learn and to change its connectivity in response to experience as it grows is much greater than that of any other species. Such postnatal neural flexibility is surely what has made cultural evolution possible. It allows our culture to provide "mind-tools" (Gregory 1981) to compensate for the time-lagged components of our brains. The biologist Richard Dawkins suggested that, because culturally acquired ideas and artefacts are reproduced by imitation and because such reproduction can also cause variation, such entities evolve by a process akin to Darwinian natural selection. He coined the term "meme" for such entities. "Memes" carried in our minds are supposed to be the controlling entities for cultural evolution as genes are for the evolution of our species. Dawkins (1976) argues that such "memetic evolution" acts on a much shorter time-scale than biological evolution and is not necessarily always to our benefit. Memes, like genes, are "selfish" in the sense that they exist only because they have been successful in being copied. "We are built as gene machines and cultured as meme machines" (ibid., p. 201). However, "we have the power to turn against our creators. We, alone on earth, can rebel against the tyranny of the selfish replicators" (p. 201).

Dawkins' idea has been enthusiastically developed by Dennett (1995) and by Blackmore (1999). I must confess that, even when developed at length, I find the idea of "memetics" as an explanation of cultural evolution to be rather muddled. It seems to me that it takes much more than a theory of behavioural imitation and selection to explain how our culture has produced Relativity Theory, Beethoven's Ninth Symphony, "Guernica," or, for that matter, Dawkins' theory of the meme. Genetic selection acts on random duplicating errors in our genes, but cultural selection acts on ideas that vary because they are modified by controlled experience and motivated thought. Nevertheless, I am prepared to accept the case that there are many components of human culture and behaviour that evolve by repeated imitation, without culture necessarily understanding how they work. One of the consequences of the theory is that the properties of memes are not necessarily those that are most useful to us, but those that cause them to be the most imitated. Such selfish components of our culture include the learned behaviour patterns that Gregory calls "mind-tools."[2]

So perhaps untidy sketching, or rather the idea of using incomplete or confusing percepts to stimulate mental imagery, is a meme that has evolved independently in different cultures. In hunter-gathering cultures natural or man-made untidy marks on rocks are used to create images for reasons already discussed. In historical times an independent evolution of the untidy sketch meme can be traced through the tradition starting with Leonardo da Vinci's advocacy of deliberate indeterminacy. "I have even seen shapes in clouds and on patchy walls which have roused me to beautiful inventions of various things, and even though such shapes totally lack finish in any single part they were yet not devoid of perfection in their gestures or other movements" (Leonardo da Vinci in McMahon 1956). The meme has since been developed by imitation and selection by many others. An example of an intermediate "meme species" is the blotting technique used by Alexander Cozens for inventing landscape compositions (Cozens 1785). Another example is the technique invented, under Leonardo's influence, by Max Ernst, of using random pencil rubbings of patterns and textures to design his surreal "frottages." He remarked, "When I closely scrutinized the sketches thus made ... I was amazed by the sudden intensification of my visionary capabilities and the hallucinatory result of the contrasting pictures" (Spies 1968). A similar tradition of using untidy images to stimulate invention evolved quite independently in the East (Rawson 1969).

Figure 7.4 shows an early sketch by Robert Venturi for the Sainsbury wing of the National Gallery in London. Clearly, a problem he had to solve was how to make the building look modern and, at the same time, to show its relationship to the existing neo-classical museum. He states that the main idea for the façade came on his second day in London as he was standing in Trafalgar Square. Whether deliberately or not, the sketch shows many of the indeterminacies advocated by Leonardo. In discussing the importance of drawing, Venturi refers to a "facility between hand and mind. Sometimes the hand does something that the eye re-interprets and you get an idea from it" (Lawson 1994). This reminds one of the finger marks made in the cave walls by prehistoric artists (Figure 7.3). Perhaps the shaman would have said, "Sometimes my hands move at random over the entrance to the underworld when suddenly by magic the spirit of an animal shows itself."

The ability to support the visualizing instinct with drawings or other incomplete percepts has never been treated as a necessary skill by our culture. It has remained a specialized meme used by artists and designers. In contrast, the memes for reading and writing are regarded today as obligatory support for the language instinct. Writing, mathematics, and other symbol systems are necessary to support the propositions, rules and reasoning that enables a technological culture to evolve. With the invention of printing, cultural evolution accelerated. However, for most of human history, the ability to read and write was the specialized skill of a minority of elite scribes. Only recently

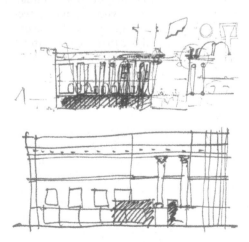

Figure 7.4 Robert Venturi, sketches (from Lawson 1994).

in Western societies have these skills become universal. It is unlikely that our brains have many innate resources for writing. Not only is the time-scale for such evolution too brief, but there has been too little selective pressure for a "writing instinct." There is no evidence that scribes produced more children than the illiterate aristocracy that employed them. As a consequence, reading and writing skills must draw on the working memory resources that evolved for visualizing. Evidence for this proposal is accumulating. Thus Brooks (1967) showed that the act of *reading* instructions interfered with a visualizing task, while *listening* to the same instructions did not. Reading can also interfere with comprehension. In an elegant series of studies Glass et al. (1985) asked their subjects either to read or to listen to high-imagery and to low-imagery sentences. They found that high-imagery sentences (e.g., "The stars on the American flag are white") take longer to verify than low-imagery sentences (e.g., "Biology is the study of living matter") when they are read, but *not* if they are heard.

Such findings should give educators pause for thought. Is it possible that the emphasis on teaching reading, writing, and arithmetic during the long period of postnatal brain growth diminishes our capacity to think with visual images? Is the postindustrial emphasis on thinking with symbols bought at the expense of thinking with images? Why do we teach all children to reason with symbols but only a few to reason with images? Perhaps one of the reasons for this imbalance in our use of cultural "mind-tools" is that we understand better how writing supports verbal thought than we understand how sketching supports visual thought. It is time to redress the balance. The ability to use untidy sketches to elicit and support our mental models is a difficult skill that we all deserve to be taught.

Description and Depiction: Interdependent Types

The cultural continuum

Twenty years ago, Palmer (1978) attempted to clarify what he considered to be muddled ideas about mental "representations," although these were (and are) central to theories of cognition. He defined a representation as an "ordered triple," consisting of a represented world, a representing world and an interpretive process that can map the first of these worlds on to the second. He then distinguished two fundamentally different systems of representation. One was the "propositional" (descriptive) system in which an arbitrary system of symbols with rules of combination (syntax) in the representing world can be mapped on to categories, propositions, and concepts in the representing world. The interpretive system must implement the rules that map the two worlds. The other was the "analogue" (depictive) system in which the representing world has varying degrees of structural similarity or isomorphism to the represented world. We can say roughly that these are "language-like" or "picture-like" representational classes. However, Palmer emphasized an interesting distinction. He termed the language-like or propositional representations "extrinsic" because the mechanisms by which they represent have to be learned and cannot be inferred from samples of the representations themselves. The other class of representations he termed "intrinsic" because, he argued, the interpretive process can be inferred from the isomorphisms within the representation itself.[3] However, there are other ways of distinguishing language-like and picture-like representations, all of which require discussion beyond the scope of this chapter (reviewed Fish 1996). Therefore, I will continue to use the vaguer but more widely understood terms "descriptive" and "depictive." Unlike descriptions, depictions are dependent on the medium of representation. Shepard (1982) has distinguished between depictions (analogue representations) with a "primary isomorphism" (as in pictures) where, for example, spatial structure is represented by similar spatial structure and colour by similar colour, with a "secondary isomorphism" in which the isomorphism is less direct. In Shepard's secondary isomorphisms, the relationship between the represented and the representing worlds is less direct, but corresponding attributes vary in corresponding ways and with similar dimensions. In contrast to descriptive systems, depictive systems cannot rely on rules of mapping but must infer representational meanings by analyzing the structure of the representation. For further discussion of these two contrasting systems, see Kosslyn (1980) and Sloman (1975).

Our culture provides us with a complex continuum of representational types that range from the very descriptive, such as language and mathematics, to the very depictive, such as film and photography or machine-generated "virtual reality." Sketches belong to a familiar class of intermediate types that possess both depictive and descriptive properties, as do maps and diagrams. Palmer pointed out that, in principle, a representational system can be infinitely extended using pointers and a hierarchical tree structure that allows descriptive and depictive systems to be combined. Thus a road map that represents space and distance depictively also contains positional descriptive symbols and these can be used as further look-up keys to descriptive or depictive information in a hierarchy of any arbitrarily determined depth. Ullman (1989) has suggested that a similar hierarchic system of depictive representations, combined with descriptive labels, might be used by the brain for object recognition.

Sketches differ from maps and diagrams in that much of the information they convey is only implicit and cannot be extracted either by analysis or by a rule system. I will argue that they are only partial representational systems that must be completed by the user's brain with which they interact. Their intermediate descriptive–depictive nature is shown by two characteristics: (1) viewer-centred depictive drawing is frequently mixed with descriptive notes and labels and (2) some of the representational elements used have both depictive and descriptive attributes. For example, the lines used to represent the silhouette contours of objects in sketches are partly descriptive. They do not exist in the pattern of luminances presented to the retinas of our eyes. Yet our visual systems seem to be able to use instinctively the information for object recognition they provide. According to a much respected model (Marr 1982), the visual system derives from the depictive image on the retina a description that can be used for comparison with stored models of objects in memory. At an early stage of processing, the brain uses a range of spatial frequency filters to extract and identify the silhouette edges of likely objects (Watt 1986). Marr calls these "the self-occluding contours" of objects. Thus the line contours found in drawings, from prehistoric times to today, probably work because they have a non-accidental correspondence with depictive to descriptive translation processes, performed by the early edge extraction stages of our visual brains (Fish 1996). Exactly how linear edge extraction works in drawings is still a subject of debate.

Representation within the brain

Clearly, cultural representational systems such as language and pictures have "interpretive systems" that depend on the brains of their users, but here the Palmer "ordered triple" runs into problems. For the physical "representing world" provided by our culture must itself be represented in the user's brain before the brain's "interpreting system" can have access to it, as also must be the "represented world." But does not this necessity lead inevitably to the much ridiculed "homunculus" paradox – an infinite regress of minds within the mind with each interpretive system needing representations of the representation it interprets? In the past this criticism has usually been reserved for the idea of depictive or "analogue" representations within the brain (e.g., Pylyshyn 1973, 1981).

To make sense, a depictive representation within the brain could only represent by Shepard's "secondary isomorphism," implemented in terms of assemblies of neurons in which connectivity and synaptic strength rather than physical structure provide the medium of representation. There are candidates for such a "secondary" depictive medium in the numerous modular regions of the visual cortex that contain arrays of columns of neurons that topographically map the visual field. Some of these have been shown by brain-imaging studies to map spatially both visual percepts and mental images (Kosslyn 1994).

However, for such neural topographic maps to qualify as the medium of representational systems (Palmer 1978) it is necessary to posit other neural processes which can scan, manipulate and interpret spatial information in ways that correspond to "design thinking." There is strong evidence that the brain provides an array of such "covert" interpretive processes to extract information from such neural representations. Moreover, this evidence seems to strengthen rather than weaken the forbidden (and ridiculed) homunculus metaphor of an inner designer, a brain within the brain, complete with an "inner pencil" to draw imagined objects (Kosslyn et al. 1988), an "inner eye" to inspect them (Kosslyn et al. 1979), "inner hands" to move and rotate them (Shepard et al. 1982) and an "inner voice" and "inner ear" (Baddeley and Lewis 1981) as the brain talks to itself.

Two classes of information, necessary for design thought within the brain, have been described as "descriptive." The first class consists of hypothetical long-term memory structures that are used both for visual recognition and mental imagery. Theories of "recognition by parts" based on contour segmentation (Hoffman and Richards 1984) or on an analysis into three-dimensional shapes (Biederman 1987) are partly descriptive. More recent evidence suggests that the brain also stores and uses individual views of objects or object prototypes which, though normalized for size and position, are nearer to depiction in the descriptive–depictive continuum (Tarr and Pinker 1989). The format of long-term remembered information about objects is important to the theory of sketch function. However, it is not clear that such information structures deserve to be classed as "representations" of the sort that can be used for design thought, for the "interpretive process" of visual recognition is an automatic retrieval and comparison process that is largely outside conscious control. However, two aspects of recognition theory are important for the understanding of designers' sketches. The first is the evidence that brains are well adapted for recognizing objects when incomplete, poorly illuminated or partly obscured stimuli are presented to the eyes. An incomplete contour fragment or object part can be used as a look-up key to long-term memory (Hoffman and Richards 1984). The second is the evidence that the long-term stored structures used for recognition are the same structures used for generating mental images (reviewed Kosslyn 1994). Recognition and imagery also appear to share resources such as those used for size scaling, mental rotation, and template matching.

The other descriptive system, language, is surely used when we reason covertly to ourselves. Experimental evidence suggests that when we reason to ourselves about categories, concepts, and propositions, we use the same components of working memory that we use when we generate or understand speech (Baddeley 1986). The existence of separate storage and processing modules as posited by the Baddeley and Hitch model for the manipulation of linguistic and visuo-spatial knowledge has recently been confirmed by brain-imaging studies (Jonides and Smith 1997).

Less well understood is the supervisory "central executive," a complex of processing resources that has been linked to planning, attention, and conscious awareness (Baddeley 1993). Perhaps a future understanding of this component will allow us to claim that there is a part of the brain, not a mind within the mind, but a cognitive recorder and controller, that can explain how it is possible for non-homunculus neural processes to create the illusion of an "inner designer" that is able to monitor, control, and report its own private information processes.[4]

Uncertainty and the need for type translation

"To invent is to choose," said Poincaré in a much quoted essay (Poincaré 1915). Design problem solving must negotiate many degrees of uncertainty. It can be thought of as the problem of choosing an acceptable route through a mental tree or bush where the trunk and branches are vague or abstract and necessarily descriptive, to some of the multiple "leaves" representing depictively concrete thought. Thus the decision tree of visual thought differs from the "problem space" of Newell and Simon (1972) in that it contains both descriptive propositions (branches) and depictive images (leaves?) with a constant need to translate between the two modes of representation. But it is a magic tree, for the act of exploration causes new branches to grow and old ones to wither.

Max Black (1937), one of the forefathers of fuzzy logic, distinguished three types of indeterminacy, all of which can be applied to sketches. *Generality* occurs when an idea that may be descriptively precise specifies a category with many exemplars. *Ambiguity* occurs when a choice has not yet been made between two or more alternatives. *Vagueness* (Black's word for a fuzzy limiting boundary) occurs whenever there is a need to specify structure, form or colour approximately for later refinement.

When we wish to represent these forms of uncertainty mentally to ourselves, we find that both description and depiction are interdependent. A visual description is useful and memorable in proportion to the number of depictive images it allows us to generate. Poetry and literature move us by their power to evoke visual memories. Since the time of Plato, it has been known that verbal ideas are easier to remember if they are associated with a visual image. The Simonides' method of "loci" by which images of well-remembered places are used to frame new material was used by the Roman orators to memorize their speeches. In his dual-coding theory of memory, Paivio (1986) has documented in detail the imagability of words in relation to meaning. A depictive image is meaningful to the extent that its inspection allows us to generate new descriptions of its subject matter. Kosslyn and others have shown repeatedly that the time to locate components of memorized spatial maps by covert mental scanning is proportional to relative distances in the actual maps. Denis and Cocude (1989) obtained similar mental scanning results when their subjects were asked to generate mental images of maps from purely verbal descriptions. We use mental images to answer verbal questions such as "How many windows does your house have?" (Kosslyn 1980). Like labelled maps, mental images are descriptively interpreted (Reisberg and Chambers 1991). Thus bidirectional translation of representational mode is a fundamental component of visual thought.

Sketching as Mental Translation

Catalysis: a metaphor

A chemical catalyst such as an enzyme enormously accelerates chemical change (new configurations of atoms within molecules) by forming a temporary compound with the reacting molecules. This temporary intermediate compound has the effect of lowering the activation energy "hill" that the reacting molecules must reach and, by temporarily holding them in the right configuration for a new molecular structure to be formed, greatly speeds up the reaction. After the reaction, the catalyst is restored to its original state. The speed is such that thousands on thousands of molecules can be transformed in a fraction of a second.

According to the metaphor, early design sketches are like catalysts in that they can combine with and transform at high speed superimposed mental information in working memory. They are not complete representations but temporary representation holding structures that help the "inner designer" to manipulate and transform the invisible representations of design thought. The mental translation "reactions" that sketch attributes catalyze are of two kinds: (1) the retrieval of implicit knowledge for depictive image generation, and (2) the manipulation and inspection of depictive images to derive new descriptive concepts. Both these mental translation processes need visual support.

Figure 7.5 shows the activation level for a hypothetical chemical reaction in which compound A-B reacts with C-D, exchanging atoms to form compounds A-D and C-B. In order for the reaction to occur, the reactants must overcome an energy barrier or hill – the activation energy. By forming a temporary intermediate compound, the catalyst (such as a biological enzyme) lowers this activation energy hill (dotted line). In the metaphor I compare mental invention to a process by which mental propositions and images are recombined in the mind in novel ways. The barrier that must be overcome is the mental effort (the amount of processing in the brain) that is needed to perceive new combinations of images or propositions. The sketch is like a catalyst in that it too forms a temporary intermediate compound with some

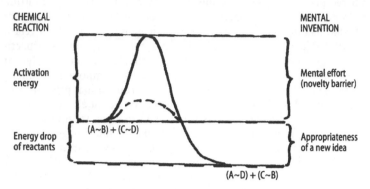

CHEMICAL CATALYSIS AND MENTAL CATALYSIS

The ~ symbol represents either chemical bonds or (by the analogy) descriptive-depictive connections

Figure 7.5 Chemical catalysis and mental catalysis.

initial propositions or mental images. This lowers mental barriers to deriving new propositions from mental images or new mental images from propositions. Just as a stile makes it easier to step over a fence by providing an intermediate platform, so, I suggest, sketches provide temporary intermediate platforms for a designer's mental representations of the task. Of course, however inventive, a design solution must fulfil its criteria. It must be appropriate. I compare the appropriateness of a new idea to that change in potential energy level of chemical reactants that drives an exothermic reaction and determines its direction.

However, perhaps this is pushing the metaphor too far. Like paintings (according to Picasso), metaphors are "lies intended to make you see the truth."

Retrieving implicit knowledge

We can create novel design concepts and objects only if we have the component parts of acquired visual expertise in memory and can retrieve these parts for fluid manipulation. "As when from the sight of a man at one time and a horse at another we conceive in our mind a Centaure" (Hobbes, 1651).

As a hunter-gathering lifestyle would demand, the innate capacity of the brain for long-term visual memory is immense. For example, Standing et al. (1970) found that after viewing 2560 colour slides for 10 seconds each, their subjects could distinguish 90% of the familiar slides from unfamiliar ones several days later. The problem for visual thought, however, is that long-term visual memory is inaccessible for manipulation. Thus retrieval capacity is a key component of visual invention that needs support in the form of verbal notes or visual stimuli.

I have proposed various mechanisms by which sketch components can provide retrieval cues for remembered objects (Fish 1996). One such mechanism is to use written notes or descriptive symbols to access, via their meaning, a complex network of associated images and descriptions. Another would be to exploit the brain's ability to use a contour fragment as a look-up key to unconscious knowledge (as when a hunted animal is partly obscured behind a bush). Hoffman and Richards (1984) suggest that the visual system decomposes object shapes into parts at regions of concavity along the visible object contour. Contour lines in sketches fit Marr's classification into: (1) contours that define an object's silhouette or surface discontinuity ("self-occluding contours"), (2) contours that mark a change in the orientation of a surface, and (3) contours that mark the edges of a shadow or change of reflectance within a surface. (In sketches lines are, of course, also used for non-contour purposes such as representing shadow and texture.) It is the first of these contours that are considered by Hoffman and Richards. According to their theory, part boundaries are defined by a rule that exploits a uniformity of nature termed "transversality." This regularity occurs because "when two arbitrarily shaped surfaces are made to interpenetrate they always meet in a contour of concave discontinuity of their tangent planes." Thus the profile of a face divides naturally into nose, lips, forehead, etc. according to this rule. The authors conclude that the visual system exploits transversality to categorize objects' parts as one of the regularities of nature that underlies the inference of parts from images. These parts can then be used for building an object description for recognition. Starting from similar premises, Biederman (1987) has developed a theory in which the input image is segmented at regions of concavity into 36 kinds of simple volumetric components such as blocks, cylinders, wedges, and cones. These are then used as primitives to derive structural descriptions of objects that are invariant for position, size, and orientation. Both these theories provide mechanisms by which incomplete sketch fragments can be exploited by the brain to elicit the retrieval of object knowledge.

As already mentioned, psychologists sometimes assess short-term memory capacity in terms of the number of "chunks" that can be handled simultaneously. A "chunk" is conceived as a meaningful structure of remembered parts that can be accessed as a whole, provided a retrieval cue with which it is associated is available. A number of studies have shown that specialized experts possess the ability to store and retrieve larger and more complex relevant memory chunks than non-experts (e.g., Chase and Simon 1973). In a series of studies, Ericsson and his colleagues have shown that mental expertise involves the manipulation of complex acquired "retrieval structures" in long-term memory. Because these can be accessed quickly from retrieval cues held in short-term memory, they allow skilled practitioners to use their long-term memory to expand the effective capacity of their working memory. These authors distinguish "short term working memory" from "long term working memory" (Ericsson and Kintsch 1995).

Reviews of expert performance have shown that, after thousands of hours of practice, experts can acquire support skills that allow them to compensate for the limited capacity of short-term memory. According to Ericsson and Delaney (1998), expert practice allows the formation of task-specific hierarchical structures in long-term memory that can be retrieved by simpler retrieval cues in short-term memory. Ericsson and Delaney showed, for example, that, with training, subjects were able to expand their memory for numbers from the normal range of about 7 digits to a level surpassing that of professional memorists (around 15–20) digits. This feat was achieved by hierarchically structuring the numbers into more easily remembered "chunks."

They have also demonstrated the skilled expansion of effective working memory in waiters (who often remember up to 16 meal orders at once without notes), chess players, and medical practitioners. There is every reason to suppose that skilled designers also have learned retrieval structures that can be accessed with shorter cues, perhaps unconscious, in short-term memory. However, this possibility is yet to be investigated experimentally.

Thus the evidence suggests that branches of the designer's descriptive-depictive tree that have been explored repeatedly will be stored in long-term memory as descriptive–depictive retrieval structures. These, when acquired, can be later accessed and manipulated by individually acquired retrieval cues in short-term memory. If this is so, then it is the brain's capacity to hold and manipulate retrieval cues to knowledge "chunks" rather than the chunks themselves that needs support.

Written notes and the untidy, incomplete contour fragments and object parts that occur in sketches are access keys to much larger memory components. These components of the sketch behave like descriptive–depictive catalysts by facilitating access to long-term visual memory retrieval structures or expert design "chunks." Improved memory retrieval facilitates the generation of a stream of depictive mental imagery. Because sketch retrieval cues contain much indeterminacy and because image generation occurs piecemeal (Kosslyn et al. 1988), this process can, in turn, generate new inventive, unexpected retrieval structures by mental recombination and synthesis.

The process proposed resembles catalysis because the semi-descriptive retrieval cues in the sketch combine temporarily in the brain with stored information during the process of remembering. A theory of retrieval by combined stimulus cue and memory trace was developed by Tulving (1983) to explain verbal episodic memory. He termed it "synergistic ecphory." In some respects, this is a visual application of his theory. Untidiness, accident, and indeterminacy amplify the inventiveness of these unconscious retrieval processes.

Manipulating and Scanning Spatial Images

In order to make the reverse translation and derive new descriptive information from depictive images, the brain provides a range of covert scanning and inspection processes. As a familiar example, suppose someone asks you, "How many windows has your house?" Most people report that, to answer this question, they take a mental walk around the rooms of their house, counting the windows as they do so. Here the inner voice and the inner eye cooperate to derive descriptive information from visual imagery. Another example is "How many corners has a sans serif, upper-case letter 'E'?" Most people report that they must mentally generate an image and count the corners to answer this question. Depictive information must be scanned to make information that is only implicit, explicit.

A series of elegant experiments on mental curve tracing (Jolicoeur et al. 1986) illustrates this. The time to determine whether or not two small marks are located on the same wiggly line is directly proportional to their distance apart on the lines, even though no eye movements are involved. An internal movement of focal attention, used to extract descriptive information, behaves as would be expected if metric distances and relations are preserved in remembered visual images. This and related work provide reminders that, although we may think of a drawing as an external optical representation, it is only available for interpretation by our brains as a preprocessed internal representation. However, in order for such descriptive processes to act on images from memory, the images need support and replacement as they fade in working memory (Kosslyn 1994). The evidence that covert mental scanning actually accesses spatially depictive images in our brains and is used to derive novel descriptive information is complex but fascinating and persuasive. It is well reviewed in Denis and Kosslyn (1999).

I have proposed a model (Fish 1996) in which the contour fragments of sparse or untidy sketches act as skeletal support structures for superimposed mental imagery. As the image is assembled, the image parts are mentally positioned, scaled, and rotated in an alignment match with parts of the sketch. If the match is successful, the image parts are then superimposed on the sketch percept as a hybrid image in which optical and image components are confused. Thus the image is sufficiently stabilized to allow mental inspection and analysis but fluid enough to support a constant stream of mental alternatives.

Hayes (1973) has provided elegant support for the use of such hybrid mental images when paper is used to support mental arithmetic. In one study he compared the abilities of European- and US-educated subjects to perform mental long division when looking at the dividend and divisor written on paper. He noticed that European-trained subjects found that such a problem was more difficult if the two-figure divisor was written to the left of the three-figure dividend (as is usual in the US and the UK, e.g., 15|673). They had been trained to place the divisor on the right (e.g., 673|15). The opposite was the case for subjects trained in the US or UK. Hayes considered two models of the solution process. In the first, the subject generates an image in the familiar format and then solves the problem using the generated (but not a hybrid) image by the usual manipulation. In the second model, the subject might solve the problem with a hybrid image, correcting the unfamiliar to the familiar format digit by digit. Thus the effects of the unfamiliar versus the familiar format would, if the first model is correct, be expected to be limited to the first stage of the solution process, while the second model predicts that the effects would be distributed digit by digit throughout the solution process. He tested this prediction with a group of 2 European and 9 American subjects each with 12 familiar and 12 unfamiliar 2-digit into 3-digit division problems. The differences in the means of subjects' times to compute successive digits of the answers for the familiar and the unfamiliar formats were distributed throughout the solution process, confirming the hybrid model.

Similarly, Chase and Simon (1973) have suggested that hybrid imagery is important in chess where the visible squares of the board support superimposed imagined movements of the pieces and, incidentally, where the existing visible pieces may interfere with their imagined new positions. However, I know of no attempts to apply hybrid image theory directly to design representation.

Convergent evidence shows that visual percepts and mental images can support each other if they are spatially compatible, but that they interfere with each other if they are incompatible (reviewed Farah 1985). I speculate that there is a selective gate between images and percepts that acts rather like the gate that controls retinal rivalry between the two eyes, ensuring that either only the most salient image is perceived or that a meaningful hybrid image is constructed. Because of the danger of image suppression by salient but incompatible visual stimuli, sketches need empty spaces for image components and must contain no distracting detail if they are to support imagery effectively. Ambiguity and incompletion in the sketch allow a flexible train of alternative images to be inspected or manipulated.

Figure 7.6 shows a sketch acting as a mental translation catalyst. Mentally represented propositions temporarily combine with written notes or symbols and partly descriptive elements of the sketch. Skilfully used, these sketch components decrease the mental effort needed to retrieve stored descriptive hierarchies and object descriptions that are in turn used to generate new depictive images. Incomplete contour fragments and other spatial components of the sketch then act as temporary holding structures for percept-image hybrids. Such hybrids provide a more stable platform for the image scanning and

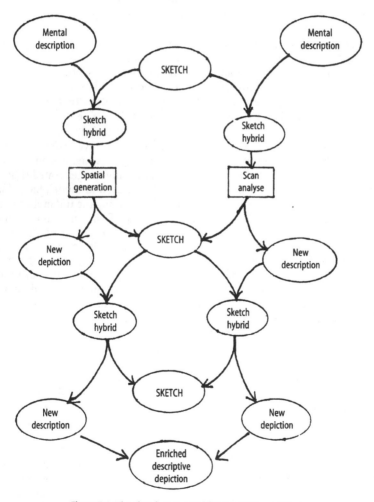

Figure 7.6 The sketch as a cognitive translation catalyst.

inspection processes needed to derive new descriptions. In the diagram in Figure 7.6, I have shown these contrasting translation processes occurring in parallel on the same sketch. It goes without saying that similar percept-memory hybrids can also facilitate descriptive-to-descriptive and image-to-image translation by accessing association memory. In her unique protocol studies of architects sketching, Gabriela Goldschmidt showed that, while sketching, architects' thought processes alternate between "seeing as" and "seeing that." Although the terminology is different, I interpret her findings as fully consistent with the cognitive descriptive to depictive translation "catalysis" metaphor described here. "The dialectics of sketching is the oscillation of arguments which brings about gradual transformation of images, ending when the designer judges that sufficient coherence has been achieved" (Goldschmidt 1991).

The Yamazaki Serving Collection: A Design Example

Robert Welch is a British silversmith and product designer who has worked a great deal in stainless steel. He tells me that he has always drawn, not only when designing, but also for its own sake. He is also a painter of quality.

Yamazaki is a Japanese company that has been involved in the production of stainless steel since 1918. In 1980 they decided to enter the American market with a range of stainless steel tableware. In 1981, after competing with four other designers, Robert Welch was selected to design a range of serving vessels – "the Yamazaki Serving Collection." The designs had to be distinguished in appearance to compete with silver and silverplate and yet be modern in feeling, for the market was believed to lie between the ages of 20 and 40. Yamazaki wanted the designer to give stainless steel a new look. The project well illustrates the use of sketches to translate descriptive constraints to depictive form. In Welch's own words, "What was needed was a brand new look for stainless steel which had nothing to do with severe forms and satin finishes" (Welch 1986). So an early generative idea was that the vessels needed a surface relief or fluting that would exploit the beautiful mirror-like reflectivity that Welch knew highly polished stainless steel could exhibit. Over 17 months of "design thought" Welch produced hundreds of drawings. Because it was not practical to make satisfactory models of most of the pieces, the drawings had not only to support Welch's own mental models of the designs but also to communicate his ideas to his client. Only two designs were ever fully modelled; the others were accepted as technical drawings only.

The design stages have been documented by Alan Crawford:

The designer and the manufacturer both had to make an act of faith in the drawings. Aware of this element of risk, Robert Welch found himself making many more drawings than he would normally have done, going over parts of the design again and again to convince himself that it would work . . . There was no question of trial runs; Robert Welch did not visit the factory. Once the technical drawings had left his studio in the idyllic little town of Chipping Campden in the Cotswolds, the design was out of his hands. During the second half of 1982, he could do nothing but wait. Then in February 1983, he drove to London to receive the first consignment of the Collection. One can imagine the nervous excitement with which he unpacked the parcels at the airport.

(Crawford 1983)

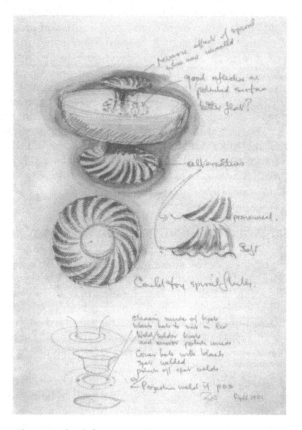

Figure 7.7 Sketch for a covered bowl (Robert Welch).

Figure 7.7 shows an early sketch for a covered bowl. It illustrates beautifully the use of a sketch to support the descriptive to depictive translation process. The spiral flutes that were finally to form such a feature of the Collection are already present, but their form is far from certain. Notice the deliberately fuzzy representation of the reflection catching the fluting and the attached written notes. One note that points to the handle of the lid reads "reverse effect of spiral when used inverted." This would surely catalyze a mental image from Welch's experience. Another reads "good reflection on polished surface . . . better flat?" Here the note preserves two possible alternatives as only an intermediate depictive–descriptive representation could. An even clearer branch point in the descriptive-depictive tree of design thought is shown by the label "alternatives" pointing to two detail sketches of fluting types "pronounced" or "soft." The sketch at the bottom of the sheet explores ways of assembling the bowl that would make it straightforward in production and easy to clean. The covered bowl did not eventually form part of the Collection. It only ever existed in Welch's mind.

Another design idea was later discarded – that of decorating the surfaces with a pattern of vertical lines (Figure 7.8). Several of Welch's drawings at this

Figure 7.8 Idea sketches for the Yamazaki Collection (Robert Welch).

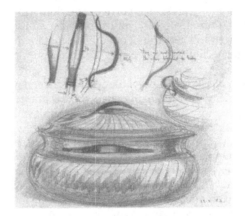

Figure 7.9 Sketch for handles (Robert Welch).

Figure 7.10 The finished Yamazaki Serving Collection (Robert Welch).

stage have a semi-hallucinatory quality as they float on top of one another as if one could see through objects superimposed. Similar hallucinatory super-impositions are found in palaeolithic cave paintings.

Although the sketch shown in Figure 7.9 is of a later stage of design thinking, there are still signs of descriptive-to-depictive translation. The note reads, "Try one end curled the other blended to body," indicating uncertainty about which of two possible types of handle to use. The tapering double-curve form was the one actually chosen and accepted, as Figure 7.10 demonstrates.

Some Implications

In this chapter I have argued that our brains evolved in an environment quite unlike today's, in which survival depended more on the ability to make quick decisions than to reflect on the long-term future. Due to this circumstance, the working memory capacities that are needed for both descriptive and depictive or visual thought are maladapted to a post-industrial culture. Our culture provides us with learned supportive behaviours to compensate for both these maladaptations. However, cultural assistance for symbolic thought is much more advanced and better understood than that for depictive thought. In general education, techniques for visualizing are the poor relations of those for symbol processing.

In a classic book, the Russian psychologist Alexander Luria documented the mind and problems of a man who had an exceptionally detailed and long-lasting visual memory, but who was unable to think with or use abstract symbols (Luria 1968). Luria outlines a number of problems of the kind that we are taught to solve with algebra that his "mnemonist" solved quickly in his head with mental images. When I tried these myself, I was astonished to find how much easier they were to solve by the imagery method than by the method I was taught. To give one example, a man and his wife go into the forest to collect mushrooms. He says to her, "If you gave me seven mushrooms I would have twice as many as you." She says to him, "And if you gave me seven mushrooms we would each have the same number." How many mushrooms did they each collect? Using the method I was taught, my working memory runs into problems. My "inner voice" runs like this: "Let H = the number collected by the husband. Let W = the number collected by the wife. Now let me see . . . $H + 7 = 2(W - 7)$ and $W + 7 = H - 7$. Er . . . Hm . . . Give me paper and pencil!" However, if instead of using symbols, I use the "inner eye" and picture to myself the husband and wife with little piles of mushrooms spread on the forest floor, the problem is strangely easy to solve without paper and pencil. I write "strangely" because the images are more complex than the symbols. Try it. The reason it is easier is that the visualizing parts of our brains are older and better adapted for handling concrete images than those for using abstract symbols. Luria remarks of his mnemonist, "His graphic images kept him from falling into the sort of errors other people can make who use formal methods to solve problems." Why was I never taught this?

It is possible that some system designers will argue that the low capacity of visual working memory and the increasing power of machine visualizing systems will eliminate the need to sketch in the early, as well as the late stages of design. Is it not better to replace our hazy mental imagery with very realistic machine imagery which can simulate in detail the light pattern presented to the retina by objects yet to exist? Inspired by James Gibson's "ecological" theory of perception (Gibson 1979) that denies the importance of mental representation, this appears to be the philosophy behind some "virtual reality" design systems (Smets 1992).

Although the usefulness of very realistic modelling systems for the late stages of design is undeniable, I believe that to discard (in machine systems) the untidy, hand-drawn indeterminacies and vagaries of "back of an envelope" sketches before we understand their function would be self-defeating. Neither is blind machine simulation of media attributes an answer. (I nickname this the "imitation bronze" approach, after those early Han period Chinese ceramics that tried to imitate bronze vessels.) Without at least a theory as to how paper sketch attributes support design, it is impossible to design appropriate machine replacements for the humble sketchbook.

I have specified elsewhere many ways in which a machine sketching system might improve on the functions of untidy paper sketches (Fish 1996). Not the least of these would be an improved facility for descriptive to depictive translation. This would include the representation of visual tolerances with machine assistance in progressive refinement as the mind explores branches and twigs of the design decision tree. Using a hierarchy of descriptive to depictive two-way pointers, machine memory can represent, below the visible surface of the sketch, a much richer and complex part of the design decision tree than is possible with older media. The evidence concerning working memory capacity suggests that sketching technology should do more than it currently does to protect our linguistic and visuo-spatial memories from trampling on each other's resources. It is easier to listen to ideas and visualize a design at the same time than to read about ideas and visualize a design. Current technology should make it easier to combine the descriptive and depictive components of our thoughts without their mutual interference.

But before such much needed descriptive to depictive translation support systems can be well designed, we need to answer questions about the philosophy of visualizing technology. For example, "Do we wish to amplify or to replace our visualizing ability with machine processing in the early stages of design?" If the cognitive catalysis theory is correct, then we cannot achieve both these objectives at the same time. Very detailed "virtual" representations would be expected to hinder rather than support the user's inventive imagery and mental translation processes. I believe cognitive amplification is a better design philosophy for sketching systems than cognitive replacement. It is in the nature of the visual arts that there must always be a part of a designer's mental image that cannot be represented by machine because it cannot be made explicit.

Another question concerns education. The evidence from prehistoric painting and sculpture, combined with the evidence from 21st-century cognitive science, shows that we cannot reason inventively, even about non-visual things, with symbols alone. Our brains have powerful resources for thinking depictively, but these can only be fully tapped if our culture can provide an appropriate visualizing technology and teach us to use it. This it has not yet done. The need arises because imagery capacity is still so closely tied to perception.

It has been known since Simonides in 500 BC (Yates 1966), that imagery can be used to augment verbal memory. However, the ability to create personal incomplete images that catalyze mental descriptive to depictive translation is a skill that deserves to be taught and practised as urgently as reading and writing. In a paper presented to a conference on "The Future of Drawing in Design" Professor Bruce Archer has argued that it is not just students of the visual arts who should be taught the art of sketching but all pupils (Archer 1997). "Drawing is indeed a great deal more than a training for the hand. It is a great deal more than a training for the eye. It is indeed a training for the mind ... On these grounds we can argue for the strengthening of its place in the National Curriculum and the confirmation of its place in entry requirements at tertiary level." Later he concludes: "I would further hypothesize that for those going on to study for any creative or inventive occupation, instruction and practice in creative vagueness should continue until at least the end of the first undergraduate year." To this bold statement I can only add that, if we are to teach all children to use more fully their under-exploited visualizing instincts, then we need to improve the quality and scope of our sketching technologies. It could be argued that the failure, as I see it, of our culture to exploit fully our innate visualizing capacity is a consequence of the emphasis that a science-based culture must give to symbolic thought. The language instinct probably evolved as an accessory to the older visualizing instinct. We have now reversed the roles of these two instincts so that mental imagery is more often perceived as an accessory to language.

If the analogy of sketches to mental catalysts is apt, then progress in the design of visualizing systems will be tied to progress in cognitive science. Sketches, it is claimed here, are not representations but representation support structures. They can only be understood by understanding the mechanisms of thought and how 30,000 years of cultural evolution have taught us to use our brains in ways for which they did not evolve. Untidy sketches provide some of the evidence. A drop of water reflects the ocean.

Acknowledgements

I am very grateful to Robert Welch for kindly giving me permission to use some of his sketches and to refer to his work in this context. I wish I had the space to do greater justice to his work.

I also wish to thank Jean Clottes, editor of Editions Seuil, for permission to reproduce two illustrations from *Les Chamans de la Prehistoire*, of which he is also a co-author.

Figure 7.4 is reprinted from Lawson (1994), by kind permission of Butterworth Heinemann, Oxford.

Notes

1. How language evolved is still a much-debated topic. For an excellent discussion of the relative roles that inheritance and culture might play in the evolution of language acquisition skills, see Deacon (1997).
2. I have necessarily oversimplified the complex arguments about the relative influences of genetics and culture on the evolution of our brains and how they work. For those readers who agree with me that this question is relevant to both how we design and how we teach the use of "mind-tools" for design thinking, the following works are recommended: Donald (1991), Durham (1991), and Mithen (1996). Each provides a different perspective on how

our brains may have both initiated and adapted to cultural change. The different theories that are presented are not necessarily mutually incompatible, however.

3. The philosopher C.S. Peirce distinguished three classes of representation which he termed "signs" (Peirce 1960). His "symbol" and "icon" correspond roughly to my use of "description" and "depiction" here. His third class is worth a note. The "index" represents by providing physical evidence of what it represents "as smoke is to fire." The most obvious example in our culture is the photograph, which is an index to the degree that it represents by sampling the light reflected by real objects. Now that machine-generated images can simulate the retinal light pattern so closely that they can hardly be distinguished from photographs, should they be termed "false indices," fiction masquerading as evidence? In a fascinating study Al Cheyne has discussed Peirce's theory of signs in relation to palaeolithic art. He too sees a psychological connection between the Ice-Age use of accidental marks on cave walls and the much more recent tradition of using deliberate indeterminacy to stimulate invention (Cheyne 1999).

4. It is often unclear in the literature exactly what is meant by "mental representation." Frequently, I suspect, the term is used to mean representation in the brain in the same sense that bit patterns in a computer can be said to "represent," say, letters and numbers. These represent not because the computer interprets them in any way, but because they are interpreted as ASCII codes by the human machine user. It is with this meaning in mind that Edelman and Tononi (2000) claim (with some reason) that memories are not representations at all but "re-entrant neural circuits" that have the capacity to repeat a given perceptual or cognitive process. According to these authors, there is no "memory code" that an external observer (i.e., a neuroscientist) can interpret as a representation. I offer a different meaning to "mental representation." I claim that thought is inexplicable unless we assume that there is a hierarchy of processes within the brain wherein higher level processes are capable of analyzing, comparing, and monitoring information "represented" by lower level processes. Thus the interpretive process is within the brain itself. The brain is mapping one part of itself on to another. Such a neural monitoring process is not a homunculus but of necessity, it must have some of the recursive properties of a "brain within the brain." How else can we count the windows in a remembered house and know what we are doing as we do it?

References

Archer, B 1997. Drawing as a tool for designers. Paper presented at The Future of Drawing in Design Conference. University of Huddersfield, School of Design Technology, U.K.

Baddeley, AD 1986. Working memory. Oxford psychology series, no 11. Oxford: Clarendon Press.

—— 1993. Working memory and conscious awareness. In Theories of Memory I, edited by AF Collins, SE Gathercole, MA Conway and PE Morris. Hilliside, NJ: Lawrence Erlbaum Associates.

Baddeley, AD and VJ Lewis 1981. Inner active processes in reading: The inner voice, the inner ear and the inner eye. In Interactive processes in reading, edited by AM Lesgold and CA Perfetti. Hillside, NJ: Lawrence Erlbaum Associates, pp 107–29.

Bauchet, R and H Stephan 1969. Éncephalisation et niveau évolutif chez les simiens. Mammalia 33:228–275.

Biederman, I 1987. Recognition by components: A theory of human image understanding. Psychological Review 94(2):115–147.

Black, M 1937. Vagueness: An exercise in logical analysis. Philosophy of Science 4:427–455.

Blackmore, S 1999. The meme machine. Oxford: Oxford University Press.

Bradshaw, JL 1997. Human evolution: A neuropsychological perspective. London: Psychology Press.

Brooks, LR 1967. The suppression of visualization by reading. Quarterly Journal of Experimental Psychology 19:289–299.

Bruner, JS and L Postman 1949. On the perception of incongruity. Journal of Personality 18:206.

Bruner, JS, L Postman, and J Rodrigues 1951. Expectation and the perception of colour. American Journal of Psychology 64:216–223.

Calvin, WH 1993. The unitary hypothesis: A common neural circuitry for novel manipulations, language, plan-ahead and throwing? In Tools, language and cognition in human evolution, edited by KR Gibson and T Ingold. Cambridge: Cambridge University Press.

Cavalli-Sforza, LL 1991. Genes, people and language. Scientific American November 1991:104–110.

Chase, WG and HA Simon 1973. The mind's eye in chess. In Visual information processing, edited by WG Chase. London: Academic Press.

Cheyne, JA 1999. Signs of consciousness: Speculations on the psychology of paleolithic graphics. Paper published on the web site, http://watarts.uwaterloo.ca/~acheyne/signcon.html

Chomsky, N 1957. Syntactic structures. The Hague: Mouton.

—— 1980. Rules and representations. New York: Columbia University Press.

Clottes, J and D Lewis-Williams 1996. Les chamans de la prehistoire. Paris: Seuil.

Corballis, MC 1991. The lopsided ape. Oxford: Oxford University Press.

Cozens, A 1785. A new method of assisting the invention in drawing original composition of landscape. London: Paddington Press. 1977 (facsimile of 1785 original).

Crawford, A 1983. The Yamazaki serving collection: Drawings by Robert Welch. Wellingborough: Skelton's Press.

Dawkins, R 1976. The selfish gene. Oxford: Oxford University Press.

—— 1982. The extended phenotype. Oxford: Oxford University Press.

De Beaune, SA 1995. Les hommes au temps de Lascaux. Paris: Hachette.

De Lumley, H 1998. L'homme premier: Prèhistoire, evolution, culture. Paris: Editions Odile Jacob.

Deacon, TW 1997. The symbolic species: The co-evolution of language and the brain. New York: WW Norton & Company.

Denis, M and M Cocude 1989. Scanning mental images generated from verbal descriptions. European Journal of Cognitive Psychology 1:293–307.

Denis, M and SM Kosslyn 1999. Scanning visual mental images: A window on the mind. Cahiers de Psychologie Cognitive/Current Psychology of Cognition, 18:409–465.

Dennett, D 1995. Darwin's dangerous idea. New York: Simon and Schuster

Donald, M 1991. Origins of the modern mind. Cambridge, MA: Harvard University Press.

Dunbar, R 1996. Grooming, gossip and the evolution of language. London: Faber and Faber.

Durham, WH 1991. Coevolution: Genes, culture and human diversity. Stanford: Stanford University Press.

Edelman, GM and G Tononi 2000. A universe of consciousness: How matter becomes imagination. New York: Basic Books.

Ericsson, KA and W Kintsch 1995. Long-term working memory. Psychological Review 102:211–245.

Ericsson, KA and PF Delaney 1998. Working memory and expert performance. In Working memory and thinking, edited by RH Logie and KJ Gilhooly. London: Psychology Press.

Farah, MJ 1985. The psychophysical evidence for a shared representational medium for mental images and percepts. Journal of Experimental Psychology General 114(1):91–103.

—— 1988. Is visual imagery really visual? Overlooked evidence from neuropsychology. Psychological Review 95(3):307–318.

Finke, RA 1980. Levels of equivalence in imagery and perception. Psychological Review 87:113–132.

—— 1985. Theories relating mental imagery to perception. Psychological Bulletin 98:236–259.

Fish, JC 1991. A model for the mental representation of sketches. Perception 19(2):277–278.

—— 1996. How sketches work: a cognitive theory for improved system design. PhD Thesis, Loughborough University, Loughborough, UK.

—— 1998. Need the system hinder your thoughts: divided attention and the design of mind-machine systems for artists. In Cybernetics and Systems '98, edited by R Trappl. Austrian Society for Cybernetic Studies, pp 215–222.

Gibson, JJ 1979. The ecological approach to visual perception. Boston: Houghton-Mifflin.

Glass, AL, DR Millen, LG Beck, and JK Eddy 1985. Representation of images in sentence verification. Journal of Memory and Language 24:442–465.

Goldschmidt, G. 1991. The dialectics of sketching. Creativity Research Journal 4 (2): 123–143.

Gombrich, EH 1966. Leonardo's method for working out compositions. In Norm and form: Studies in the art of the Renaissance. Oxford: Phaidon Press.

Gregory, RL 1981.Mind in science: A history of explanations in psychology and physics. Cambridge: Cambridge University Press.

Hayes, JR 1973. On the function of visual imagery in elementary mathematics. In Visual information processing, edited by WG Chase. London: Academic Press.

Hobbes, Thomas ([1651] 1968} Leviahon. London: Penguin Books.

Hoffman, DD and M Richards 1984. Parts of recognition. Cognition 18:65–96.

Jerison, HJ 1991. Brain size and the evolution of mind. Fifty-ninth James Arthur Lecture, American Museum of Natural History.

Jolicoeur, P, S Ullman, and M Mackay 1986. Curve tracing: a possible basic operation in the perception of spatial relations. Memory and Cognition 14:129–140.

Jonides J and EE Smith 1997. The architecture of working memory. In Cognitive neuroscience, edited by MD Rugg. London: Psychology Press.

Kosslyn, SM 1980. Image and mind. Cambridge, MA: MIT. Press.

—— 1994. Image and brain: The resolution of the imagery debate. Cambridge, MA: MIT Press.

Kosslyn Sm, Ball, TM and Reiser, BJ 1979. Visual images preserve metric spatial information: Evidence from studies of image scanning. Journal of Experimental Pschychology: Human Perception and Performance, 4: 47–60.

Kosslyn, SM, CB Cave, D Provost, and S Von Gierke 1988. Sequential processes in image generation. Cognitive Psychology 20:319–343.

Kosslyn, SM, WL Thompson, and NM Alpert 1997. Neural systems shared by visual imagery and visual perception: a positron emission study. Neuroimage 6:320–334.

Kosslyn, SM and AL Sussman 1994. Roles of imagery in perception: Or, is there is no such thing as immaculate perception? In The cognitive neurosciences, edited by M.S. Gazzaniga2. Cambridge MA: MIT Press, pp 1035–1104.

Kosslyn, SM, WL Thompson, IJ Kim, and NM Alpert 1995. Topographical representations of mental images in primary visual cortex. Nature 378:496–498.

Kosslyn, SM, A Pascual-Leone, O Felician, S Camposano, JP Keenan, WL Thompson, G Ganis, KE Sukel, and NM Alpert 1999. The role of area 17 in visual imagery: convergent evidence from PET and rTMS. Science 284:167–170.

Laitman, JT 1983. The anatomy of human speech. Natural History August 20–27.

Lawson, B 1994. Design in mind. Oxford: Butterworth Architecture.

Leakey, R 1994 The origin of Humankind. London: Phoenix Press.

Leakey, R and R Lewin 1992. Origins reconsidered London: Abacus Books.

Lewin, R 1993. The origin of modern humans. New York: Scientific American Library.

Logie, RH 1995. Visuo-spatial working memory. Hillside, NJ: Lawrence Erlbaum Associates.

Logie, RH and KJ Gilhooly 1998. Working memory and thinking. London: Psychology Press.

Lorblanchet, M 1995. Les grottes ornés de la prehistoire. Paris: Editions Errance.

—— 1999. La naissance de l'art. Paris: Editions Errance.

Luria, AR 1968. The mind of a mnemonist. Translated by Lynn Solotaroff. Cambridge, MA: Harvard University Press.

McMahon, AP, ed. 1956. Leonardo da Vinci's treatise on painting. Princeton, NJ: Princeton University Press. (This is a facsimile of 16th-century transcriptions by an unknown copier of Leonardo's notes now in the Vatican library. Most of the original Leonardo manuscripts have been lost and so cannot be dated exactly. Comparison with autobiographical Leonardo manuscripts that have survived suggests that the transcription is faithful.)

Makirinne-Crofts P, J Fish, S Sadaat, and W Godwin.1992. Improving computer systems for fashion designers. Report for the Science and Engineering Research Council. UK.

Marr, David 1982. Vision A computational approach. WH Freeman.

Minsky, M 1985 The society of mind. New York: Simon and Schuster.

Mithen, S 1996. The prehistory of the mind: The cognitive origins of art and science. London: Thames and Hudson.

Newell, A and HA Simon 1972. Human problem solving. Upper Saddle River, NJ: Prentice-Hall.

Paivio, A 1986. Mental representations: A dual coding approach. Oxford: Oxford University Press.

Palmer, S 1978. Fundamental aspects of cognitive representation. In Cognition and Categorization, edited by E Rosch and B Lloyd. Hillside, NJ: Lawrence Erlbaum Associates.

Peirce, CS 1960. Collected papers of Charles Sanders Peirce. Volume 2. Cambridge MA: Harvard University Press.

Phillips, WA and DFM Christie 1977. Components of visual memory. Quarterly Journal of Experimental Psychology 29:117–133.

Pinker, S 1994. The language instinct. New York: Penguin Books.

Poincaré, H 1915. Mathematical creation. In The foundations of science, translated by GB Halsted. Pennsylvania: Science Press.

Pylyshyn, ZW 1973. What the mind's eye tells the mind's brain: a critique of mental imagery. Psychology Bulletin, 80, pp. 1–24.

—— 1981. The imagery debate: analogue media versus tacit knowledge. Psychological Review 88(1):16–45.

Rawson, P 1969. The appreciation of the arts/3 drawing. London: Oxford University Press.

Reisberg, D and D Chambers 1991. Neither pictures nor propositions: what can we learn from a mental image? Canadian Journal of Psychology 45(3):336–352.

Segal, SJ 1972 Assimilation of a stimulus in the construction of an image: The Perky Effect revisited. In The function and nature of imagery, edited by PW Sheehan. New York: Academic Press.

Segal, SJ and S Nathan 1964. The Perky Effect: incorporation of an external stimulus into an imagery experience under placebo and control conditions. Perceptual Motor Skills 18:385–395.

Shepard, RN and LA Cooper 1982. Mental images and their transformations. Cambridge, MA: MIT University Press.

Shepard, RN 1982. Perceptual and analogical bases of cognition. In Perspectives of mental representation, edited by J Mehler, ECT Walker, and M Garrett. Hillside, NJ: Lawrence Erlbaum Associates.

Sloman, Aaron 1975. Afterthoughts on analogical representation. In: Theoretical issues in natural language processing, edited by RC Schank and BL Nash-Webber. Arlington, US: Tinlap Press.

Smets, G 1992. The theory of direct perception and CAD in virtual reality. Paper presented at Virtual Representations for Design and Manufacture Symposium. Coventry University.

Spies, W 1968. Max Ernst frottages. London: Thames and Hudson (revised edition).

Standing, LG, J Conezio, and N Haber 1970 Perception and memory for pictures: single-trial learning of 2500 visual stimuli. Psychonomic Science 19:73–74.

Tarr, MJ and S Pinker 1989. Mental rotation and orientation-dependence in shape recognition. Cognitive Psychology 21:233–282.

Tuluing, Endel 1983. Elements of episodic memory, Oxford psychology series, 2. Oxford: Clarendon Press.

Ullman, S 1989. Aligning pictorial descriptions: an approach to object recognition. Cognition 32:193–254.

Watt, R 1986. Visual processing: computational, psychophysical and cognitive research. Hillside, NY: Lawrence Erlbaum Associates.

Welch, R 1986. Hand and machine. Chipping Campden: Welch.

White, R 1999. Prehistoire. Sud Ouest: Paris.

Williams, GC 1966. Adaptation and natural selection: A critique of some current evolutionary thought. Princeton, NJ: Princeton University Press.

Wexler, M, SM Kosslyn, and A Berthoz 1998. Motor processes in mental rotation. Cognition 68:77–84.

Yates, FA 1966. The art of memory. London: Routledge and Kegan Paul.

Zeki, S 1993. A vision of the brain. Oxford: Blackwell Scientific Publications.

Ziman, J 1976. The force of knowledge. Cambridge: Cambridge University Press.

8

The Thoughtful Mark Maker – Representational Design Skills in the Post-information Age

Martin Woolley

Skills – Past, Present And Future

Skill, technology, and design

The impact of computing confronts all professions, profoundly affecting the way that they function and challenging occupational relevance and survival. A common misperception is that emerging technologies automatically replace traditional skills, making many specialist skills redundant, effectively deskilling expertise. This chapter explores deskilling in the context of design representation and discusses the potential of new technologies as both a generator and absorber requiring new techniques, approaches, and thinking with which to create a new skills-base.

The application and definition of traditional skills are discussed, with a comparison of the differences that have taken place in the development of design practice between knowledge-based skills and the separation from implementation. This is followed by an analysis of the representational skill-base currently in use in design practice, together with comment on the effects of technological change. Lastly, strategies to develop appropriate representational skills for the future are proposed.

Skills through history

E.P. Thomson describes and evokes an era when the hand drove production and an individual's skills were a complex continuum of manual dexterity, a keen eye, knowledge of the task /role of the object in use, and often direct knowledge of their "clients":

The skilled workman is taught by his materials, and their resources and qualities enter through his hand and thence to his mind. The artefact takes its form from the functions which it must perform, the "dish" of a wheel from the movement of the horses, the ruts in the tracks, the weight of an average load. These are not finely calculated on paper, they are learned through practice.

(Thomson 1993, p. xi)

Throughout history, skills have been delineated by socio-economics; the acknowledgement of skills was rooted in the language: "having a trade," "serving an apprenticeship," "having a vocation," and "being qualified." The application of the terms "skilled," "semi-skilled," and "unskilled" effectively delineated and defined a social caste system. Pre-industrial skills were predominantly manual with tangible outcomes. As machines replaced manual dexterity, unskilled workers became expendable. Skills became less visible and, with the advent of machine intelligence, eventually dispensable. Today, only advanced skills and knowledge are appropriate for maintaining sophisticated automated production. To achieve these skills, the growth of a fully equipped learning environment, rather than ill-equipped teaching, is increasingly viewed as a necessity.

Definitions of skill are imprecise, for, like "design," the word has been overused. Significantly, the obsolete meaning of skill is 'understanding'; however, skill is more commonly used to define a trade or technique, requiring special training or manual proficiency. McCullough acutely observed and accurately analyzed skill as being:

Acquired by demonstration and sharpened by practice. Although it comes from habitual activity, it is not purely mechanical. This is evident not only in the fact that some skills are difficult to measure, but also in that skill can become the basis of a vocation.

(McCullough 1996, p. 3)

The schism in skills occurred during the Industrial Revolution because it disconnected design from production and origination from making. Many interconnected skills became redundant as the mechanisms of mass production first separated out the differing levels of skill, then began their systematic eradication, from low-level manual to advanced non-manual skills. More recently, many areas of skilled origination are no longer required professionally. In parallel, the general regard for skills diminished with the advent of "commercial idealism" engendering "better products for all." As a result, current definitions of skill tend to be narrow; typically, they are focused on subtasks, special abilities, and training, linked to learned ability with acquired knowledge. The holistic skills of making and doing, conceiving and realizing as a continuum, are consequently rarely addressed. The personal ownership of an integrated repertoire of "hands-on" skills used to make objects has been replaced by a fragmented, externally referenced skill-base.

In contrast to and separated from making artefacts, design skills had previously been focused on concept origination, representation, and productionization. Many of the varied skills previously associated with both designing and making became the exclusive skills of design, dominated by ideas translated through manual skills – drawing, rendering, and model-making – combined with the creative/inventive skills of information gathering and problem solving. Design skills became less visible and not automatically discernible in the final manufactured product.

Design became increasingly separated from consumption; however, as Plante (1997, p. 77) observes, the differences between designer, consumer, user, and used are not clear-cut. For example, the deskilling of production may be mirrored by the deskilling of associated end-user tasks, as in the case of automated camera production and function. Conversely, production deskilling can result in the effective transfer of skills from professional to end-user. The do-it-yourself market is a manifestation of this transference, where many traditional construction or making skills have been replaced by less demanding and more flexible installation and assembly skills.

Skills – keeping ahead professionally, socially and psychologically

Little theoretical analysis of skills occurred until comparatively late in the post-war period of industrial development, when attempts were made to define them in relation to professional certification, resulting in an artificial division between professional (white collar) and non-professional (blue collar and usually manual) skills. To distinguish clearly between the knowledge differences within the skill-base of each, the former were deemed to be knowledge-based; the latter were deemed implementation skills.

In both groups, skills varied from habitual, unconscious reflexes performed repetitively – e.g., riding a bycicle or playing a keyboard, to high-order skills associated with specialist knowledge requiring concentration and risk assessment – e.g., performing surgery or manipulating advanced representational software. The key factor is practice context. Habitual skills are appropriate for repeated consistent application; high-order skills respond to complexity, changing conditions, unpredictability, and are frequently associated with risk.

Computers perform an infinite range of tasks and software developers have attempted to translate high order into habitual skills, regulating tools to make actions predictable, repeatable, and reversible – for example, the use of a "paintbrush" in computer-aided design (CAD) software no longer requires an understanding of paint viscosity. Commonality is increasing with the use of standard interface symbols which allow previously unique operations to be identified generically across a range of software. The flexibility and variety of control options makes it impossible to compartmentalize computer skills, for, like the skills deployed in playing the guitar, they operate at almost every skill level. Customized operating systems are becoming ever more intelligent, adapting to the user's competencies in all but the high-order end of the spectrum.

Initially, most representational technologies deskill, simplifying control and removing the unpredictable. There is a cascade effect with the decreasing cost of data processing impacting design functions, as high-order tools in a few practised hands become habitual tools in the semi-skilled hands of the many (Figure 8.1). Once deskilled sufficiently by data processing, the average person can learn many design skills "playing" with the software, developing skills seamlessly and acquiring simple habitual skills en route. The extensive capabilities of current representational software mean that the trained user is unlikely to acquire skills in all functions, particularly as, unlike many manual tools, their technological counterparts generally lack integration with one another. This lack of truly integrated design tools poses difficulties, particularly in the initial development of a design: the sketch and concept model early stage.

Figure 8.1 The diffusion of representation technologies into wider ownership.

While the power of data processing has brought about increasing efficiencies, particularly in the repetitive or mechanistic aspects of design, the removal of the unpredictable is closely associated with the loss of intimate interaction. The removal of creative serendipity by data processing bypasses and therefore ignores the levels of skill that determine the difference between the "unhappy accident" and the creative breakthrough. Indeed, it might be argued that highly refined or perfected skills militate against such abrupt advances by limiting the exposure of the skilled operative to fresh, if accidental, possibilities. In addition, making the tool, as well as using it, can be seen as extending this exposure to new stimuli by multiplying the expressive variables through which skills are applied. Furthermore, directly experiencing and reflecting on the representations made with the tool can also be seen as reinforcing the knowledge of the tool's known and unknown potential. This knowledge can ultimately determine the form, performance, and appearance of the resultant artefact in use.

Exercising coherent and integrated manual skills within the triad of tool making, tool use, and tool outcomes was a powerful creative force within the pre-industrial world that might be assumed to have little contemporary relevance. However, there are unifying parallels in design communications and computing that suggest that the potential already exists for similar creative gains in contemporary professional practice if required:

- *Tool making* equivalents exist in the reprogramming or reconfiguration of software controls and the construction of highly adaptable, manually operated, data input devices.
- Skilled *tool use* occurs during as in the operation of advanced design software controls in combination with such devices.
- *Tool outcomes* occur through engagement with the virtual artefact via increasingly vivid representational output devices or through the more tangible results of rapid prototyping.

Such realistic forms of representation arguably constitute "intermediate products" in which the barriers between non-working prototype, working prototype, and end product are dissolved by digital means. This process is complete in areas such as multimedia.

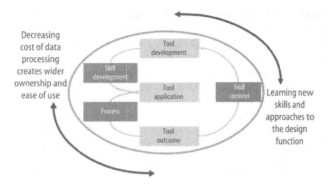

Figure 8.2 Reiteration – representational tool feedback loops inform and empower the designer.

Within this context of continuous data-processing systems, the potential for powerful feedback loops exists throughout the entire design, production, and distribution process and is particularly significant in the relationship between tool production, application, and outcome required by the just-in-time and concurrent engineering approaches to manufacturing (Figure 8.2). Such feedback systems already occur quite readily, but are commonly unstructured and unrecognized within contemporary design representation. The uptake of intelligent systems provides the potential for new "tools for tools." Such communication and analytical devices could provide information systematically, simultaneously improving the tool, its skilled use, and indirectly contributing to the personal development of the practitioner.

Seemingly, computer evolution continuously delivers increased power, efficiency, and ease of use, while simultaneously withdrawing difficult, unpredictable, and costly skills from many areas of professional life. The expertise used for "serious work" has been diverted to "non-serious," low-risk contexts. However, this scenario fails to take account of the continuing invention and refinement of representational tools and environments at the high end. Such advances require not only new skills, but a strategic vision that facilitates their acquisition throughout the professional working life of the designer.

This process is characterized by the gradual transfer of advanced traditional representational tools into wider ownership, through cost reduction and the increased ease of use of related hardware and software systems. Desktop publishing, for example, demonstrates how high-order typographic and graphic processes have been simplified and disseminated into wider, semi-skilled ownership. In this context the employable design practitioner maintains a dominant paid position only through a combination of originality of work and its representation, together with access to new representational technologies and wider involvement in the design/production process.

Sketching tools for the skilful: concurrency and changes in design protocol

Skill represents a paradox in human terms: the requirement for it often reduces the accessibility of practice by the wider, unskilled population, while skilled practice can generate unique personal and collective motivation. This symbiotic relationship between skill and creativity is often overlooked and the basic human need to acquire skills is underestimated. Clearly, skills are valued within the employment market, but their relationship to long-term personal and professional fulfilment and development is not so obviously addressed. It is clear, however, that professions that fail to regenerate skills continuously may atrophy.

Skills can be linked to overcoming imperfections or constraints, both environmental and human. One manifestation of this linkage is the way in which skill engages with the unpredictable. Different occurrences, including the irregularities inherent in natural materials such as wood, have traditionally introduced unpredictable elements into creative processes; this can be both stimulating and frustrating. Managing recurring risk and the serendipity associated with design sometimes assumes mystical proportions: a studio potter is not in full control when pots are "committed to the flames"; their final quality is unknown until they emerge from the kiln. The special relationship between the skilful and the random can be found in such diverse activities as improvisational jazz or rock climbing. In design, serendipity occurs throughout all the traditional representational processes, from inconsistencies inherent in materials to the inaccuracies of hand/eye coordination.

The transition from pencil to marker-pen systems has affected the risks associated with skill both positively and negatively. Marker pens facilitated, for the first time, the accurate replication of intense colour, requiring new manual skills to control tonal application. Similarly, the computerized selection and specification of colour has increased accuracy and repeatability, but the new functional links between colour, texture, and tone require significant new control skills to progress beyond basic applications.

The issue of risk versus control can be seen most clearly in the sketch representation, where the fluidity of developmental thinking and construction parallels the open-ended flow of conceptual thinking in the early stages of the design process. In this case approximation is the most significant characteristic of sketching, alongside a tolerable and possibly stimulating level of hand/eye unpredictability. In the same context the drawing is often worked through selective iterations as a simultaneous inductive/deductive process in which mappings, evaluations, and discourses gradually evolve together as design synthesis.

Conversely, accuracy is the most telling feature of formal technical drawing and clearly explains how such drawing readily transfers to CAD – parametric systems demand numerical absolutes, which are perfect for zeros and ones but not for charcoal. Sketch designs can be made in either two or three dimensions and at one extreme make the invisible visible – i.e., make tangible the resolved concept held in the designer's mind. At the other extreme, they allow the designer to begin without preconceptions and, by exploiting the vagaries of hand/eye coordination, to model an unanticipated representation, combining simultaneous thought and action, effectively "making on one's feet."

A sketch design often blends text with drawing, creating annotation that allows the designer to manipulate visual form with contextual information. The rapid convergence technologies of speech recognition and approximate visual representation means that, for the first time, simultaneous real-time annotation and computerized drawing should soon be possible. This potential should not be underestimated, as it will allow the designer to contextualize, visualize and make decisions within a single coherent communication process. It will change the use of the sketch from an essential but discontinuous process, to the status of a readily transmissible resource carrying information forward to further, more accurate representational stages, as part of an evolutionary flow of data. Integrating continuous reflection with design practice adds a new dimension to the "reflective practitioner" model developed by Schön (1998, p. 21), pointing the way to the accelerating pace of individual enquiry at the root of much experimental design. Coyne further suggests that:

All enquiry begins with engagement. Tensions and irresolutions require response. Reflection can be defined as the process of going outside the immediate situation – "to something else to get a leverage for understanding it" – and involves the search for an appropriate tool. The tool is part of the active productive skill brought to bear on the situation. The tools that feature in the reorganization of the experience include theories, proposals, recommended methods, and courses of action. The applicability of the tool is worked out in the situation.

(Coyne 1995, p. 39)

In this context, continuous representational data flow will present new opportunities for reflective commentary which both overlay and increasingly intertwine with visualization and perhaps serve as a timely reminder that text is as significant a modelling material as clay or graphite.

The elements of the design process – initial ideas, concept generation, research development, testing, and prototyping – have become part of a continuous data flow that facilitates productionization: moving seamlessly from conceptual development and prototyping, to manufacture, using a single evolving data resource. The continuous transmission of representational data that enables design development phases to run concurrently provides several advantages. This transmission has implications for design beyond speed and increased flexibility, greater fluidity and risk reduction, and marks a profound change in the fundamentals of design practice. This change in design protocol is potentially as significant as the change that occurred when design separated from production.

The control of skills

The transition of skill throughout history, from the development of man the toolmaker to the evolution of data processing, suggests changes that are both intrinsic and extrinsic, and indicates an evolving personalization of skill choices and organization, engendered by both socio-economic and technological forces. The development and use of representational skills closely parallel this historical transition, which is identified in Figure 8.3. The degree of individual control over the acquisition and exercising of design/making skills since the early industrial period has shifted from prescription to autonomy, and suggests a further development whereby the skills of leading-edge practitioners might be largely self-determined.

The control of tools and the development of skills can be traced through four generic phases:

1. *"Prescribed or regulated" skills* were commonly associated with early mass production when requisite skills were specified in great detail and

uniformly developed through apprenticeship training in order to fill the more complex gaps in incomplete automation. This situation occurred primarily because creativity and origination skills were viewed as direct threats to uniform quality control.

2. *"Evolved" skills* were associated with traditional crafts where techniques and tools evolved over generations, primarily in response to the inherent difficulties of fabrication by hand and, to a lesser degree, by the vagaries of demand. A degree of self-selection with regard to skilled technique was often tempered by practical tradition, with creativity and origination skills being largely confined to the perceived qualities of workmanship and the interpretation of traditional requirements.

3. *"Self-directed" skills* are addressed by the autonomous contemporary designer/maker whose learned skills are controlled and redirected under individual control, with creativity and origination skills prioritized and integrated within technical and making skills.

4. *"Self-originated" skill* applies to broad design data processing and has little historical or evolutionary precedent, since broadening diversity and functionality demands new skills of a high order to accommodate continuously evolving practices. Such skills cannot be "handed down"; they require the practitioner to become proactive in their origination, development, and ownership, with creativity and origination skills prioritized in all aspects of technical, design, and skills development.

Significantly, it is the amalgamation of self-directed and originated skills that is most likely to influence the emerging "cybercrafts" driven equally by designer-producers and designer-makers.

In terms of the training required for many computer-assisted modelling systems, the acquisition of appropriate interactive skills is usually overlooked

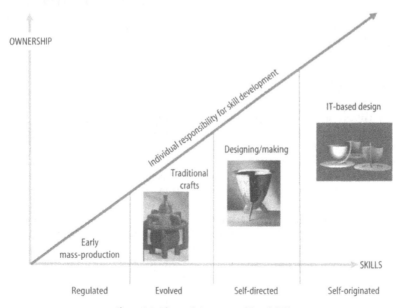

Figure 8.3 The evolving ownership of skills.

in favour of the simple memorizing of control options. Thus the responsibility for the definition and development of skills is effectively passed on to the user and such skills become, by definition, self-originated. This transfer of responsibility can have unexpected benefits in allowing the skill-conscious practitioner to seek out novel ways of exploiting or combining control options. In this context, it is suggested later that specific personal aptitudes, plus fresh forms of lateral thinking, can be developed to provide a foundation for reskilling.

Representational Futures

Training for life: "ubiquitous computing" and professional development

A word processor's primary value is that it provides the option for the text to "evolve" non-sequentially, replacing the linear process that is required in traditional writing. There are comparative similarities in subcomponents of the design process; however, the lack of integrated systems inherent in computer modelling has so far largely prevented continuous data processing and its associated benefits. The position becomes clearer when tracking the recent history of computing. The development of computers began with the first-phase mainframe revolution which took place in the 1940s and 1950s; the second phase, which occurred in the 1970s, was the PC era; the current third phase, "ubiquitous computing," commonly denotes an explosion in the numbers of computers, other intelligent devices and computer connectivity. Design representation has become a more fluid, continuous parallel to the design and development process, rather than a series of milestone snapshots. In design terms, this transition is increasingly allowing representational processes to reflect continuously and parallel a fused mature design and production system (Figure 8.4). The notion that specialized, hands-on peripheral devices might be linked to advanced representational technologies is demonstrated in the increasing number of highly skilled professional fields – for example, the training for, and practice of, medical surgery.

The extension of sensitive control peripherals beyond the mouse and screen pen is evolving at a rapid pace in fields as diverse as music, computer games, and surgery and is slowly gaining momentum in design. With the emergence of such sophisticated control extensions, the common view that the simplistic virtual reality environment will be the next natural extension of the design toolbox should be reappraised. In contrast, the notion of an augmented reality could be considered, in which control peripherals bridge the gap between hand, eye, and machine intelligence. These peripherals would be combined with viewing tools that allow the "real" world to be blended with the virtual, and doing tools that integrate the electronic hand implement with the haptic feedback device.

Clearly, design and production are fusing, driven by the malleability of data. Tools that exemplify this fusion include initially crude, but increasingly adaptable, "sketching" systems that are linked directly to rapid prototyping equipment and integrated manufacturing systems. This connectivity enables fundamental design changes to be made throughout the duration of the

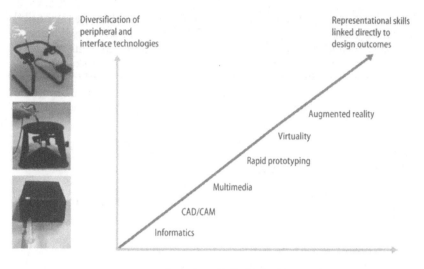

Figure 8.4 The increasing continuity of design processing has facilitated the development of skill-based tools.

project, a crucial factor when responding to a volatile consumer market, as evidenced by stereo-lithography which allows some aspects of modelling to become a form of productionization. The "rules" that govern manufacture can be in-built for the duration of the project, informing all modelling processes. Rapid prototyping used in tandem with production detailing permits a conceptual and contextual view to be retained throughout the process.

Evidence in support of this progression can be found in a recent doctoral thesis by Yi-Luen Do (1998, p. 6). A "Right-Tool-Right-Time" prototype program demonstrates how a freehand sketching system that infers intentions would support the automatic activation of different design tools based on a designer's drawing acts. Such an example identifies the interactive duality of augmented reality in that it both interprets and informs representation, while simultaneously allowing real world tools to provide the primary data input. Thus an augmented reality would both simulate and interact with a tangible reality in order to provide enhanced cognitive and control potentials. Not that the skills associated with such systems would necessarily remain apparent, as Mark Weiser, former head of the Computer Science Laboratory, Xerox Park, suggests: "Disappearance is a fundamental consequence not of technology but of human psychology. Whenever people learn something sufficiently well, they cease to be aware of it" (Weiser 1995, p. 78).

Despite this evidence, such an augmented reality, which deals effectively with skill-as-creative stimulus, has yet to become available. In the meantime, although skill is an essential factor in the human psyche and an important component of the design process, professional development lacks coherence. Computing has not deskilled design but created confusion; skills have become isolated and short-term, difficult to maintain and develop because of the pace of new system introduction. To prepare for innovative and unpredictable skill development, while reinterpreting the traditional skill base with regard to ubiquitous computing, the design community should, where possible:

1. Participate in the development of augmented reality representational tools through alliances with the appropriate developmental industries.
2. Continuously re-skill in relation to these evolving tools.
3. Consider the use of such tools as conceptual drivers and not merely as facilitators.

The designer's indemnity, and claim to professionalism, is to adopt the successive technological developments ahead of their diffusion to the wider public. The necessary reskilling of designers will take place within economic volatility: employment is increasingly characterized by part-time, discontinuous work, self-employment, subcontracting and multi-skilling. Adaptability and flexibility are crucial to survival in all professions, and the skills required by the practitioner will be the ability to discern the transferable core, and the relevance and use of leading-edge technologies. To stay ahead, practitioners should move towards an augmented reality, be capable of continuously reconfiguring attitudes to skill to exploit further innovation, and develop a more proactive realization of tools. Proactive practitioners should address:

- The active pursuit of technologies that directly connect artificial intelligence with physical dexterity.
- Continuous rather than segmented design data processing.
- The deployment of technologies for individual/organizational learning and representational skills acquisition.
- The implementation of comprehensive, coherent virtual, and tangible working environments.
- Design alliances that identify new computational tools and develop new methods of use/application.

Developing a body of work that bridges virtual, augmented, and tangible realities is fundamental to the optimization of personal skills (Figure 8.5), a process that has much in common with the systematic exploration of materials, processes, tools, images, and finishes practised by designer-makers. "Systematic play" could aptly describe this form of design learning, allowing multiple media to record a personal body of work recording and analyzing the applications and outcomes. A significant aspect of an augmented reality environment encompassing design representation would be integrated analytical tools that allow personal development to be tracked through the augmented reality equivalents of sketchbooks, notebooks, sample collections, and test pieces. This aspect would form an effective new basis for Schön's (1998, p. 157) reflective practitioner model and is echoed in the aims of advanced industrial training: "Integrating state-of-the-art rapid product development technologies with fundamental master workmanship" (Industrial Centre 2000, p. 1).

This integrated approach supports the development of tools by, rather than for, designers. Investigation into the nature of design skills will provide constructive guidelines for developing representational tools and applications for the emerging technologies, including haptic feedback devices which redefine body/eye/brain skills. The most useful transferable skills facilitate the transfer of the design process between tangible and virtual worlds, with the partial

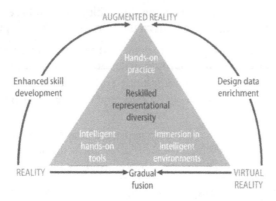

AUGMENTED REALITY

Enhanced skill
development

Design data
enrichment

Hands-on
practice

Reskilled
representational
diversity

Intelligent
hands-on
tools

Immersion in
intelligent
environments

REALITY

Gradual
fusion

VIRTUAL
REALITY

Figure 8.5 The high performance studio.

rather than total immersion of the designer within a digitized environment being more relevant to the changing design process.

To imbue an object with visible quality requires an effective relationship between tools, methods, and skills. A symbiosis between these factors is more difficult to maintain in data processing; representational machine intelligence is able to present a perfect surface image, a holistic vision of high quality design, and, as systems become more sophisticated, this potential increases. However, such "perfection" ultimately represents orthodoxy imposed by the software industry, rather than meaningful design choice. Practitioners should seek visualization tools that allow genuine exploration and run counter to this orthodoxy. If reflecting on design is "like looking at the film of a car crash run backwards," associated representational technologies should possess the facility to re-create barely controlled chaos. The availability of such systems will not come about by waiting: progress inevitably requires coaxing, including lobbying by both design practitioners and theorists to form creative alliances with technology developers.

The design practitioners' industry should be equipped with both transferable and specialized representational skills to fulfil this alliance and reflect the value placed on broader intellectual capital. The attributes of positive action and ownership within the emerging technologies should be considered essential and include the ability to conduct personal design technology audits and provide a related specification. Technology is moving so quickly, becoming so diverse, that designers need to become involved in the structural make-up of representational systems, not merely their deployment. Knowledge that allows designers to determine the form, as well as to take control of their new tools, will make them more flexible and support the continuous creativity of their work.

In this context, a frequently overlooked limitation of current IT training for designers concerns the customization of hardware and software. Practitioners exploit their equipment resources, but are not necessarily responsible for their specification. A prerequisite of advanced skill in this context is understanding, responsibility and bonding, achieving an intimate relationship between the applications and the tools. With traditional skills, knowledge was handed down from craftsman to apprentice, the user inevitably customizing tools. Sturt (1993, p. 134) records the mix of tool purchase and adaptation that were common in the blacksmith's trade. Skills were associated with a repetitive task rather than demanding fresh insights and techniques. The master craftsman possessed a high degree of control over his tools – their selection, origins, and final form; engendering a parallel mindset in the computer-based design field would be desirable to develop skills in applying, specifying, and customizing IT.

Representation technologies continue to move forward, providing increasing performance, while more design tasks are paralleled within emerging technologies and associated specialist peripherals. Future designers require the ability to discern the underlying principles of technological progress, focusing this ability on their own personal and strategic development. Thus there is a recurring requirement for design practitioners to take charge, to possess a working contextual knowledge of their representational systems and an understanding of the development cycles of technologies within a professional context, not merely to exploit developments once they are established. Without this facility, no matter how skilled, creative or able the designer, there will be an increasing risk of technological redundancy. Clearly, this risk also applies to many other professional groups and their equipment requirements, but the understanding and control, customizing development/adaptation of the technologies and tools is particularly significant for the designer.

Computer-aided representation in transition

As previously argued, a coherent reskilling philosophy requires transferable knowledge of, and attitudes to, skill; the acquisition of such a philosophy might be encouraged on the basis of life-long learning for design practitioners. McCullough (1996, p. 22) observes that the emergence of computation as a medium, rather than a set of tools, suggests a growing correspondence between digital work and traditional craft. He points to the rapid improvements in the physical relationship between users, computers, and the external actions they perform, particularly new forms of direct physical control and haptic feedback. This relationship implies an increasingly sensitive relationship between man and machine: where there are barriers/linkages between direct thought, vision, and intention, the transmission of thought into action is indivisible. Manifestations of this relationship lie in free-form areas of dance and singing, where body and mind are united in expressive and emotionally charged variants. McCullough makes a strong case for a similar level of technological and human integration; the requisite skills become those of hand (or body) and eye, mirroring the traditional crafts. The field of studio ceramics is a useful example in this context; soft clay malleability parallels the instant malleability of virtual digital "materials." Eye, hand, brain become an integrated production mechanism in which "skill" is denoted by the potter's ability to envisage and create simultaneously, with little conscious awareness of the formal, physical interventions required. The example of ceramics is further echoed in the way that humans are able to physically merge with sophisticated haptic controls.

Figure 8.6 illustrates how traditional representational practices have evolved in the post-information era through the changing relationships between *organization* (the management, processing or organization of

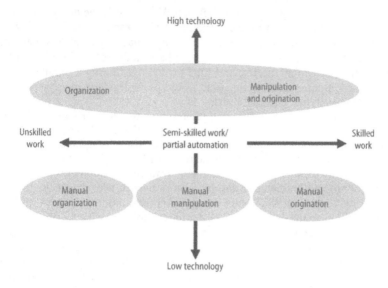

Figure 8.6 The positioning of skill in relation to representational technologies.

completed representations), *manipulation* (representation through the manipulation of pre-existing entities, and *origination* (the creation of representations without deploying pre-existing entities). Technology has integrated all three and, in particular, has closely engaged origination with manipulation. Although this process of integration has, to a degree, deskilled origination, it can be argued that new skills have evolved in relation to the manipulation of entities and that both are now better supported by integrated, highly automated organizational capabilities.

How successfully this harmonization has resulted in improved decision-making or representation is open to question, for, as Tenner (1996) has stated, "The problem is that there is growing evidence that software doesn't necessarily improve decision-making" (ibid., p. 204). In terms of "augmented reality" tools (i.e., tools that require the application of "real-world" skills within virtual design environments), this suggests that more effort is required for the selection and design of HCI hardware to ensure a closer fit between skill potential and machine options.

The intelligent marker or sharp pencil

There is no means of future proofing design representation, given the vagaries of technological development and the uncertainties of market demand. Rather than displacing skills, evolution requires the recombination of human, social, and professional requirements. Ideally, future strategies should be based on near-certainties: the technological context, increasing power of information technology and production cost reduction, will require increasingly sophisticated skills; to stop the new race of robots from ruling the world (Warwick 1997, p. 32).

A barrier to such strategies is represented by the over-functional nature of computer software design applications, because software has evolved by incremental addition rather than through unified multifunctional tools. Like the Japanese kitchen knife, the skilled professional requires flexible multifunctional tools rather than the fragmented gadgetry that is prevalent in the Western kitchen. Such a shift would result in a continuity of control, compatible with the increased continuity of design data flow; a traditional markmaker can be used in a variety of ways. Tools should respond to and dictate skills, creating new forms of representation whose controls literally lie at our fingertips. Given the predicted growth of haptic feedback devices, a truly "intelligent" interactive "pencil," together with other similarly transformed design tools, should soon be possible. Such devices are an increasing necessity if designers are to address the processing of continuous data.

At this point in time, design practitioners might look enviously at music where, partly as a result of the digital basis of harmonic systems and annotation, it has long been possible to input data across a wide range of instrumental interfaces. These range from the traditional (using matching and often hard-won skills) to the innovative (requiring new or often no skills). Representational instruments have few parallels: there is little outside the computer that can mimic traditional drawing skills, while simultaneously exploiting new processing options. Perhaps design practitioners and theorists alike are partly to blame for not providing sufficient market-pull to harness appropriate levels of technology-push. This missing interface has meant that a generation of designers has been relatively weak at building directly on their traditional skill-base and has had to adapt to the machine, rather than harness existing hand/eye skills to machine intelligence.

The notion of an intelligent, intellectually "sharp" pencil or "thoughtful marker" might serve as a useful technological metaphor and ambition. Such an instrument would possess an interpretative, processing and information function, in addition to its primary data input role. It would respond to skills without boundaries, allowing the user to perfect hand/eye/brain routines through successive iterations of technological advancement. Such transferable representational skills linked to both personal and corporate knowledge would be relatively future-proofed.

Norman (1998, p. 221) argued convincingly for a more human-centred approach to the design of computers, primarily as a means of making their potential more realizable and accessible to ordinary consumers. As part of this thesis, advanced markets are characterized as immature, their products difficult to learn and operate, and their development driven by technologists. Such products inevitably demand high-level skills, but frequently these might be characterized as *purposeless skills*, i.e., those that essentially attempt to overcome the inadequacies of the system in contrast to *purposeful skills*, which link high-level human abilities to positive outcomes (Figure 8.7). Clearly, these two types of skill are not always easy to separate and the former may occasionally become the catalyst for the latter. Nevertheless, Norman's dictum could be applied to professional augmented reality technologies within the marketplace and not merely to their volume-distributed, consumer progeny. In this way, augmented reality tools may overtake their rather ineffective and deskilled virtual equivalents, although, as Norman argues, this process may require corporate as well as user action, so that the designers and the eventual users of the products interact. In this case, the eventual users are

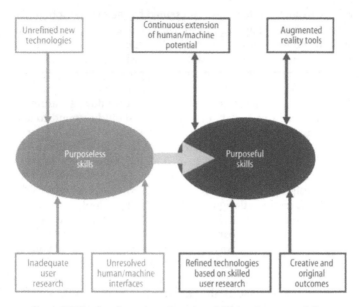

Figure 8.7 The transformation of transient skills into long-term skills.

designers so that cooperation might be assumed to be more, rather than less, fruitful than their designer/consumer equivalents.

In conclusion, several factors point the way forward: first, designers should be equipped with the specification skills required to take more control over new working environments. The design profession – through associations, professional groups, education and research projects – should influence/press the IT industries to develop technologies that adapt to existing high-level skills and new skills, providing software systems with truly integrated, responsive design tools rather than add-on functions.

Second, new models of representation should be developed which address continuous design data processing and practitioners should become more active participants in the development of appropriate design data-processing tools. Finally, as the profession prepares for a period in which real and artificial worlds collide, it should ensure that the positive contribution that each might make to a purposeful reality is not lost in a predetermined and exclusively technology-driven race to acquire totally immersive skills. The starting point of these three seminal activities should be the cultivation of increased knowledge about skills, including their past, present, and future value. Design practitioners should emerge from such deliberations equipped with purposeful skills that have been acquired partly as a result of, rather than a denial of, their supposedly "purposeless" counterparts. Such an approach would represent a true, self-initiated "training for life" (Industrial Centre 2000, p. 1).

References

Coyne, R 1995. Designing information technology in the postmodern age – From method to metaphor. Cambridge MA: MIT Press.

Industrial Centre (The Learning Factory) 2000. Prospectus. Hong Kong: The Hong Kong Polytechnic University.

McCullough, M 1996. Abstracting craft: The practiced digital hand. Cambridge MA: MIT Press.

Norman, D 1998. The invisible computer: Why good products can fail, the personal computer is so complex and information appliances are the solution. Cambridge MA: MIT Press, 1998.

Plante, S 1997. Zeros & ones: Digital women & the new technoculture. London: Fourth Estate.

Schön, D 1998. The reflective practitioner: How professionals think in action. UK: Ashgate Publishing.

Sturt, G 1993. The wheelwright's shop. Cambridge, UK: Cambridge University Press.

Tenner, E 1996. Why things bite back – Predicting the problems of progress. London: Fourth Estate.

Thomson, EP 1993. Foreword to The wheelwright's shop, by George Sturt. Cambridge UK: Cambridge University Press.

Warwick, K 1997. March of the machines: Why the new race of robots will rule the world. London: Century.

Weiser, M 1995. The computer in the 21st century. New York: Scientific American Books.

Yi-Luen Do, E 1998. The right tool at the right time: Investigation of freehand drawing as an interface to knowledge based design tools. PhD thesis, Georgia Institute of Technology.

9

Design Representation: Private Process, Public Image

Gabriela Goldschmidt

Dedicated to the memory of Tom Heath, who was an architect, scholar, educator, and wonderful person.

Introduction

In a lovely little book titled *What, if Anything, is an Architect?* the late Tom Heath (1991) offers no definition, and no direct reply to the question evoked by the title. Rather, the book is a collection of short articles about issues that pertain to architecture and to the architect's activities and concerns. We would like to adopt a similar strategy. At the outset we should be asking: what (if anything) is design representation? Instead, we shall briefly outline some of the underlying dimensions that we think are of importance when considering design representation, architectural or otherwise.

Designers represent – and design representations are made – before, during, and after the process of designing any entity, regardless of whether the designed entity is being constructed, manufactured, or assembled as a "real" product. In fact, the ultimate design goal is to arrive at a satisfying representation of the designed entity: bringing the "real" entity into being is a task that usually falls into realms other than design, and actors other than designers (e.g., builders and manufacturers) are responsible for it. We may argue that to design is to represent, and in no case is there design without representation.

Representations are not all of a kind: on the contrary, representations differ vastly in purpose, in modality, in the media they use, and in their level of abstraction (Grignon 2000). Representations may be internal – in the mind – or external, i.e., material and physically perceivable. The former are of great significance in the mental processes of reasoning about one's design, but it is the latter that we shall address in this chapter. Representations are the basis for communication among team members in a collaborative design effort (e.g., Benaïssa and Pousin 1999), whereas in the case of an individual designer, they facilitate the dialogue of the designer with him- or herself and with the design materials (Schön and Wiggins 1992). Some representations are elaborate, precise, and detailed descriptions of the designed entity, whereas others are "quickies," rough outlines of initial ideas. Representations may be concrete or abstract, true to scale or lacking scale altogether. They may be pictorial, written or otherwise expressed in a language of symbols. Some are three-dimensional, like scale models of buildings, but most are two-dimensional and consist of marks on paper (or computer monitor). They may adhere to conventions, such as the rules of perspective, for example, or they may be free interpretations of the designed entity.

Representations vary in consistency: they may give a full and detailed account of all parts and all aspects of the designed entity; at other times, they may be partial, pertaining to selected elements only, or displaying different components with varying amount of detail and attention (Herbert 1988). Some representations are vague and depict a general concept only, and others are not concerned with the physical properties of the designed entity but with operational properties that are best expressed by diagrams.

It is of great interest to study design representations from the standpoint of the designer or designers: of the almost endless possibilities, what kinds of representation do they choose to make at various phases of a design process? Why are some types of representation and modes of representing privileged? How typical are certain representational characteristics to particular designers, or to kinds of design tasks, or to a domain of design? How situated are such representational modes in historical and cultural contexts? And, equally important, we should also shift our attention to the receiving end: what impact do design representations have on those for whom they are intended? Are representations influential and, if so, in what manner?

In this chapter we propose a framework for the study of design representation along two perpendicular axes: the axis of the image, with its private and public poles, and the axis of epistemological dimensions, i.e., the intellectual realms in which a discussion of design representation is relevant and timely. These dimensions are *cognition, history and culture*, and *technology and media*. We shall discuss the questions regarding private and public images within each of these three dimensions.

Cognition

Design problems are ill-structured and it is therefore necessary to conduct a search en route to a design solution since there are no problem-solving algorithms as in the case of well-structured problems. A design search is primarily aimed at eliciting potent preliminary ideas, a design concept that can be developed and refined into a concrete solution proposal. In the process of generating, developing, and assessing ideas, one *reasons* about them: the designer or design team inspects ideas and images, sources of inspiration, partial solutions, and so on to ensure their relevancy, their congruence with requirements and constraints, and their "good fit" with one another. This is how the potential contribution of alternative courses of action to a possible solution is established. The process of reasoning requires that the ideas in question be represented, so that one can react to them, transform them, refine them, or reject them. Since our cognitive apparatuses are endowed with the capacity for mental imagery, we make extensive use of it in design reasoning: visual imagery is the locus of inner representations in which designers elicit and entertain design configurations. However, despite its powerful affordances imagery is restricted by many factors such as the sharpness of the image, its duration, and our ability to read information off it and manipulate it. External representation is meant to compensate for the limitations of inner representation, and this is why the individual designer resorts to external representation as of a very early phase of the design process. When a designer works with others or reports to them, external representation is mandatory, of course – communication depends on it; but the dialogue the designer conducts with him- or herself, is no less significant.

For the most part, external design representations take the form of drawings, photographs, models, and other artefacts. The drawing is the primary medium of representation, starting with rapid preliminary sketches, advancing to so-called "hard line drawings" and "presentation drawings," and finally technical construction or manufacturing drawings. From the point of view of cognition, rapid freehand sketches are of the greatest interest to us, as they are the closest evidence to the designer's mental processes that we have access to. Fish (chapter 7 of this volume) argues that visual imagery has evolved in humans to assist in hunting, and is geared at fulfilling tasks like pattern recognition and differentiation. Designing complex artefacts or buildings poses very different representational requirements, ones that imagery cannot cope with without external support. Fish stresses that, in terms of evolution, imaging capacities have hardly had the time to adapt to the kinds of requirements that designing imposes. Sketching, in this view, is the easiest and most immediate external device that we can use to amplify mental imagery and to extend it (Fish and Scrivener 1990), and it is well suited to the task of designing new entities. Goldschmidt (e.g., 1991) holds a similar view regarding the relationship between imagery and sketching; she proposed that internal and external representations establish a feedback loop in the course of reasoning about forms and configurations in designing. She has therefore referred to the combined representational activity, in the mind and on paper, as "interactive imagery."

Sketching is, indeed, the most effective, cognitively economical, and rapid means of experimentation at the disposal of the experienced designer (unskilled sketchers benefit much less from the use of sketching in their design search. See, for example, Verstijnen et al. 1998). Sketching is used not just to document what has taken shape in the designer's mind, but also to actually generate form and shape. Shapes on paper can be transformed and retransformed; they can be worked out with more or less detail; they can change size and location within seconds. Moreover, even random marks on paper can be seen as harbouring cues for design ideas and configurations, and sketches are therefore a medium through which discovery and invention are facilitated (Schank-Smith 2000; Suwa et al. 1999). The individual designer who sketches in order to support his or her own thinking may use a personal "shorthand" that permits sketches to remain vague and abstract, which in turn allows him or her to stay non-committal – an advantage in the early phases of the design process. Sketches can be made in short or long sequences, thus encouraging search for as long as necessary. If need be, sketches can be easily discarded, as the effort involved in their making is modest, thus assisting in the making of a fresh start when one is called for. Sketching is, therefore, most economical in terms of cognitive goal-oriented activity, which explains its wide and universally extensive use in designing (Goldschmidt 2002).

It is therefore not surprising that designers (as well as artists) started making sketches as soon as a suitable medium – paper – had become readily available and affordable, in the last quarter of the 15th century. Figure 9.1 reproduces a sketch by Leonardo da Vinci, describing a design for a movable bridge for military purposes (c. 1480). Leonardo's sketch is explorative and informal; it depicts several ideas and includes notes. It is quite safe to say that we sketch in a similar mode to this very day at the conceptual phase of designing.

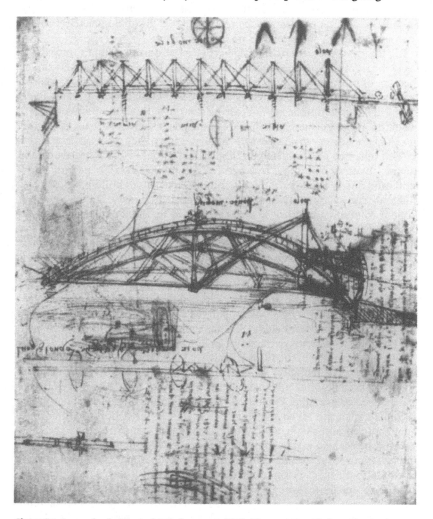

Figure 9.1 Leonardo da Vinci, sketch for a movable bridge, c. 1480. (*Codex Atlanticus*). From *Leonardo da Vinci: Engineer and architect*, edited by P. Galluzi 1987. Reproduced with permission of the Montreal Museum of Fine Arts.

To date, the freehand sketch has no substitute in the private process of shaping design concepts. Reliance on sketching varies among design domains and is definitely dependent on the type of task and on individual propensities. But to the extent that the individual designer reasons about his or her design, sketching is how he or she carries out a dialogue with the design situation (Schön 1983).

In teamwork things happen a little differently, of course, as reasoning is carried out collectively. On top of serving the individual's needs, representations are also the basis for communication among team members who use language to explicate their design arguments and make them accessible to others. Gestures and artefacts that are used to model or simulate shapes of design components or their mode of operation also facilitate the flow of ideas and arguments concerning the proposed design's appropriateness. As a rule, cognitive economy is the overriding principle that dictates the use of the fastest, simplest, most direct and least effort-exerting mode of representation for a given task and under specific circumstances. The process of designing involves the production of sequential representations, until a "satisficing" (to use Herbert Simon's term) solution is reached. Once a candidate solution is proposed, there is no substitute for a suitable drawing or a 3-D representation that specifies what words, for example, can only approximate. The design team therefore uses representational means similar to the ones in use by individuals, but, due to an increased need for communication, verbal representation in particular plays an increased role in this type of design activity.

When inspecting the cognitive aspects of the public image of a work of design, we all but switch paradigms: here it is not the designer's cognitive mechanisms that are of interest. Instead, it is the user, or viewer, who occupies centre-stage, and it is his or her perception of the representation that we focus on. In successful cases the user infers information that facilitates proper understanding or use of the designed object. Principles of perception, such as those devised by Gestalt psychology, are therefore included in design textbooks (e.g., Quarante 1994) as a prerequisite to the understanding of issues of human factors. On a practical level this means representations that facilitate the use (or production) of a designed entity with no operational mistakes or unnecessary waste of time or energy. If that entity is a building, we are talking about easy orientation and way-finding, coherent hierarchy of spaces, and generally speaking a minimum need for signs and instructions. Bamberger and Schön (Schön 1983) have coined the term "felt-path" to describe how the viewer "experiences" a building when only its plans are available. Based on those plans, the viewer is able to follow a virtual path through the building's spaces and "feel" what it would be like to move through them. Yates (chapter 1 of this volume) gives an eloquent example of the use Le Corbusier made of photographs of houses he designed from specific angles to achieve the desired viewer's impression of the design.

But there is also a cultural level at which the representation is to convey the status of the building or product and the proper attitude towards it. This is done by employing explicit and implicit symbols and by using norms and conventions through which messages are communicated. Attributes related to the approach to a building, the height of its major spaces, the manner in which daylight is admitted into it – all speak for formality or lack thereof, for quotidian or ceremonial functions, and so on. Likewise, in typography, composition may easily help to distinguish, at first sight, an ordinary printed text and a text of special significance, such as a religious text. We ascribe values such as "status" or "sophistication" to certain types of representation (e.g., specific fonts), and we can create alternative representations of the same product (e.g., a consumer product) to give it "leisure" or "office" looks. Designers are aware of users' perceptions and prepare their representations accordingly. For example, a frontal perspective will always yield a more formal-looking appearance than a top-down axonometric view of the same building. What is selected for presentation and how it is represented will always affect the viewer's notion of a work of design.

History and Culture

As is still the case today in indigenous architecture, traditional architecture of the past was produced by designers who worked primarily according to the dictums of convention, habit, and common sense. Innovation, creativity, and even individuality were not held in prime esteem. Alexander (1964) calls such design "unselfconscious" and he contrasts it with today's "selfconscious" design, produced by designers who have little, if any, commitment to tradition, convention, or habit. On the contrary, the contemporary designer is expected to boast originality and creativity above all else and, if successful, he or she is rewarded by peers and the general public alike. However, no designer can "reinvent the wheel" with every new design. One therefore works within a style, one "subscribes" to shared values of a design culture or microculture that is the product of historical circumstances, and one accepts rules and regulations imposed by authorities, such as safety codes. This is true for the individual designer and, even more so, for the design team, which can hardly function without some agreement on what its members aspire to achieve and how they are to proceed about it. The affiliation with a microculture also impacts representational choices, which help to solidify and communicate an individual or organizational identity. In the context of today's global economy, design is a major factor in the creation of distinct images of products or lines of products. Corporate identity, designed to enhance the competitive edge of companies, is based on the creation of a unified representational "language." The choice of such a "language" is clearly an act of positioning oneself on a socio-cultural "map," one that best fits the market niche for which the products are destined.·

It is therefore not surprising that designers develop personal or communal (in a design firm, say) "trademarks" that eventually differentiate them from other designers and contribute to the ability to identify their work. We would like to suggest that architects, and to a lesser degree also industrial designers, develop personal repertoires of shapes and forms, as well as rules of assembly and composition of these forms. In teamwork, the more prominent designers usually establish these repertoires that are then shared by coworkers, who in turn are in a position to influence and develop them. A designer's repertoire is dynamic and may undergo many changes over time, although at any given period (the length of which varies) it is often quite stable and fixed. The talented designer manipulates his or her repertoire in endless ways so that each resultant design is unique, but the basic repertoire may be quite limited. Let us look at two examples. The first is Alvar Aalto with his "fan motif" (Quantrill 1983). Aalto was a keen sketcher and in many of his sketches we notice that, when first approaching a new design task, he often sketched a scheme that had the shape of a fan. We suggest that Aalto retrieved a dominant image from his repertoire, a fan-like shape, and represented it as a starting point for a new design because he knew that for him, this shape is readily manipulatable. Indeed, the shape was then worked and reworked systematically and only many a sketch later was Aalto finished with the process of transforming and changing what eventually yielded a new design. In his next project, however, he was just as likely to again start with the fan motif that would lead him on a different design journey each time.

The second example, also from the realm of architecture, concerns the "kit of tools" used by the architect Mario Botta. In an interview with the curator of an exhibition of his work, Botta was asked about the dominance of axiality in his plans (Wrede 1986). He replied that he had often wanted to break with axiality, but it seems stronger than his will and keeps coming back, to Botta's own surprise. There is no ideological reason for the pre-eminence of the axis in Botta's work. Rather, it is one of the strongholds of his compositional repertoire, a principle that is flexible enough to support many different design solutions for various types of buildings. We often find circles superimposed on an axis in Botta's plans, practically independent of the size and type of building in question. Figure 9.2 shows plans for two very different buildings – a crafts centre and a private house – both based on axial circular plans. According to Trevisiol (1982) "All Botta's work is directed towards the continual refinement of a number of simple, elementary forms" (ibid., p. 82). To actually refine those simple forms, Botta has to draw them time and again, transforming them as he progresses from one sketch to the next. Like Aalto, Botta imposes a representational device that creates the kind of order that he can work with and develop. We postulate that both Aalto and Botta use representational means to organize the materials of a new design problem such that those materials would readily fit into the context and format that the designers feel comfortable with, and for which their competences are hewn. Thus representation plays the role of an important adjustment tool that helps to transfer specific problem situations into the cultural (or micro-cultural) design world in which the designer can confidently act.

Not every designer practises a somewhat obsessive work-mode of the kind we discern in the cases of Aalto or Botta, but designers do have personal representational preferences. Personal preferences pertain to the amount and nature of sketching (whole versus detail, small versus large scale, abstraction and incompleteness, quality of line, repetition and over-tracing, the use of written notes, and so on). Bugaisen (2000) has demonstrated that architects can be shown to use different characteristic patterns of representation in terms of the use of types (diagrams, orthogonal projections, axonometric and perspective views) and categories (free, overall and detail sketch, and hardline) of drawings. Such preferences can be plotted into a matrix that reflects their frequency; over several projects a personal "profile of preferences" of an architect emerges. Interestingly, designers of the same generation, or those who share the same ideologies, do not necessarily have similar representational preferences. There is, however, some evidence that designers whose built works of architecture share common traits also tend to have similar profiles of representational preferences (ibid.). This interesting finding requires further study.

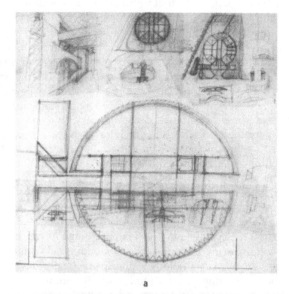

a

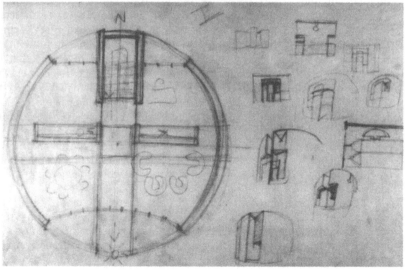

b

Figure 9.2 Mario Botta, preliminary studies. **a** Crafts centre, Balerna, Switzerland, 1979. **b** First-floor plan of private house, Stabio, Switzerland, 1980. Reproduced with the permission of Mario Botta.

This is the place to stress that, despite historical variety in representational modes and preferences among designers, all graphic design representation is essentially based on the system of orthogonal (or parallel) projections that was developed in Italy at the beginning of the 16th century. This development is attributed to Raphael who needed an analytical method of representation to solve the complex design problems he was faced with in his practice.

Alberti's "perspectival projections" which had been published some 80 years earlier (following Brunelleschi's invention of the principles of perspective) served as the basis for the new orthogonal projections which we use, with very few changes, to this very day. A close look will reveal that the history of visualization through projections is an intricate story that poses a great many questions, such as those eloquently evoked by Evans in a remarkable study (1995). However, at a simple, operational level, it is the solidity of this common base of all but the entire body of graphic representation in design that makes it possible to carry out valid comparisons among representations of different individuals or groups.

At the hands of the individual designer, representation fuels the private design search process, which is inevitable at the outset of a new task. When the design is solidified, its representation ceases to be a private matter and it takes on a public nature: when presenting to colleagues, juries, clients, or the public at large, one aims at gaining approval. To this end, one must choose the representational strategy that is best suited for the messages one hopes to convey to a target audience. When laborious water-washed drawings were prepared in the Ecole des Beaux Arts, the students knew that there were strict representational norms that they had to observe in order to succeed, and they spent most of their time training to master these norms. In modern times norms are less strict, and participants in competitions, for example, are instructed as to what information their drawings must include, although normally they are not told how they are to present the required information. As we have seen, norms do develop as a function of professional trends and under the influence of wider cultural realms like the arts, and as a result of scientific and technological developments. Cultural and social conventions, then, determine to a large degree the kinds of images that designers endeavour to construct and put in the public eye through representation.

It is most interesting to inspect periods of cultural shifts in which old norms of representation appear inadequate. In the 20th century we encounter two such periods: the 1920s, with the birth of Modernism, and the 1970s, when postmodernism largely replaced it. Klevitsky (1997) has shown how, alongside new representational means like the collage, the relief or the proun, which were basically two-dimensional, abstract three-dimensional compositions were also introduced into design training offered by two avant-garde schools of the 1920s, the Bauhaus and the Vkhutemas. These compositions were made of readily available materials like wood, metal, fabric, cardboard, glass, and so on – all materials that could also be used for the construction of product or architectural models. The three-dimensional abstract compositions, however, were not models. Rather, Klevitsky (ibid.) refers to them as a special kind of "three-dimensional sketches", not quite sculptural but definitely in search of compositional distinction. Many of those compositions had a very dynamic character. Students who exercised representation in all two- and three-dimensional media were expected to have broadened their conceptual horizons while also expanding their representational capacities, so as to better explore and express new conceptual design potentials.

In the 1970s the new spirit of postmodernism that invaded architecture brought new design paradigms and agendas to theory and practice alike. In practice, some of the major innovations included a desire for richness of form to the point of compositional dissonance, exaggerated forms, independence of the building envelope, and a taste for classicism. Design theory strived to liberate architecture from abstraction and over-functionalism. Instead, the creation of an architecture of "narrative contents" became a leading concept (Klotz 1988). State-of-the-art work was concerned with telling a story at least as much as it was concerned with solving design problems; parallel

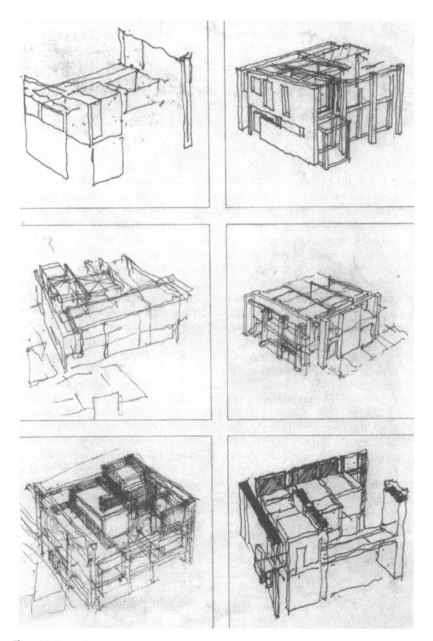

Figure 9.3 Peter Eisenman, published sketches for "Houses of Cards," 1980s (from Eisenman 1986, p. 23). Reproduced with the permission of Oxford University Press.

developments can be detected in the arts (conceptual art, for example, flourished in the 1970s). Architects were interested in concepts such as the "memory of the site" (Grumbach 1979); this was not easily representable within existing conventions and architects sought to expand them. Unlike their Modernist predecessors of the 1920s, the postmodernists did not invent new representational means but took new liberties with existing ones. For example, they revived the use of axonometric drawings and stretched them to the limits of coherent expression. The office of Stirling and Wilford started making "worm's-eye" (up-view) drawings; in these drawings only selected elements of the designed building were shown – just enough to document an overriding idea, a central concept. A good example is the publication in major European architectural magazines of three competition entries for museums in Germany by Stirling and Wilford, between 1975 and 1977. In addition to many such "axos," and even more surprising for contemporary readers, these

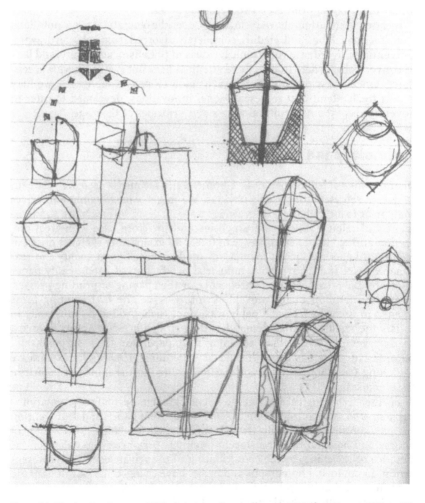

Figure 9.4 Charles Gwathmey, published sketches for a coffee-pot, 1990 (from Tapert 1990, p. 52). Reproduced with the permission of Charles Gwathmey and Rizzoli International.

publications also included a large number of small preliminary sketches (see Chapter 2). Architects have, of course, been making sketches in abundance for centuries, but very rarely have they been published, especially as competition entries. Stirling pioneered a practice that has since become very common – the publication of rough preliminary sketches. These sketches, which had until then been private, have become part of the projected public image of a work of design and even somewhat of a trademark of the designer, and we find them in design and architectural publications as of the second half of the 1970s. Figure 9.3 shows an excerpt from Eisenman's sketches for "Houses of Cards." In a book he published on the theme (Eisenman 1986), a 44-page chapter is devoted entirely to sketches; many of them, as well as hard-line drawings elsewhere in the book, are axonometric views. Figure 9.4 reproduces sketches by Gwathmey for a coffee-pot, made on a lined writing-pad sheet. These sketches, too, were deemed appropriate for book publication (Tapert 1990). It is hardly imaginable that both these images would have found their way to print even two decades earlier.

In later years attempts were made to use idiosyncratic design notations (e.g., Tschumi, Libeskind, Hadid), completely divorced from habitual drawing conventions, in order to communicate ideas of pluralism, ambiguity, and lack of conventional order. To date these attempts have not (yet) won wide acceptance. Once a new major cultural wave again swipes at our shores, we may expect fresh attempts at representational innovations, developed to cater to novel needs of the new cultural values and priorities of their age.

Technology and Media

That the state of technological development and the media used are pertinent to design representation seems obvious. We might want to remind ourselves that, prior to the 15th century, drawings were made on surfaces such as parchment and vellum, which were expensive and therefore used with great discretion and economy. The practice of freehand exploratory sketches does not begin until after the "paper revolution" made paper of good quality and reasonable price readily available. Industrial production of good quality paper at affordable prices was the consequence of the opening of printing presses, after the invention of the movable printing type. The first printing press opened in Rome in 1467 and paper was, since then, produced industrially to satisfy the rapidly growing demand of the printing presses. Artists were fast to discover the outstanding advantages of paper for their purposes and became ardent consumers of this old-new medium, primarily for explorative sketching (see Figure 9.1), which was well suited to the innovative spirit of the Renaissance.

Further technological advances that had a great impact on representational capacities were the introduction of semi-transparent paper, and the invention of light-sensitive chemically treated paper, used for "blue-prints" and, of course, in photography.[1] Translucent tracing paper once again enhanced the practice of sketching, thereby contributing to the private aspect of representation. In contrast, photographs and prints have changed the way designs are communicated publicly, be it for technical purposes (blue prints) or as artistic interpretations of works of design, for which photography is frequently employed. With the advent of computers and various modelling, drawing, and rendering applications, the question of technology and media in design representation has gained new significance, and representational skills, and their acquisition, have attained an importance of a new order of magnitude.

Skill ownership appears to be the major issue for the individual designer in his private design enterprise. Woolley (Chapter 8 of this volume), who talks about the deskilling of contemporary designers as a result of greater automation in design and its representation, sees a need for reskilling. However, whereas in the past designers acquired skills that were developed by others, i.e., traditional or borrowed techniques, Woolley thinks that reskilling is contingent on the designer's involvement in developing design tools. Accordingly, we may expect breakthroughs in Computer Aided Design (CAD) to depend on the participation of designers in software developments.

A case in point is the state of the art of computational support tools for the conceptual phase of designing, which lags behind other CAD applications that have already had a strong impact on subsequent stages of the design process (the production of technical documents, evaluation, presentation, and more). Current technological limitations are not the only reason for the slow development of support for conceptual design. To succeed, support must address basic cognitive operations of designers (the conceptual phase of designing is by and large a private affair, whether carried out individually or by a team). The introduction of affordable paper, and later of tracing paper, revolutionized the process of designing precisely because it allowed on-the-spot experimentation and representation cycles that are affordable both in terms of human resources (time and energy involved in sketching), and in terms of material cost (of paper). Computer applications that fail to "understand" that directness and immediacy of representation are of the essence in the generation of design ideas, have little chance to support natural design behaviour. It has been demonstrated time and again that the natural attributes of design ideation are most robust, and attempts to ignore, circumvent, or change them have thus far yielded disappointing results. It is therefore estimated that only when designers themselves take responsibility for the development of the computational representation tools they wish to put to work for them, can we expect real progress in this domain. In addition to effectively direct the growth of the offspring of today's CAD (and virtual reality), modelling and simulating, managing data-bases, case libraries, precedents and other computational design support tools must be further perfected. Facilities for collaboration, including at a distance, should allow team design, which is the order of the day. We should also take into account that new technological possibilities always redefine design itself. The relative ease with which we can today animate representations, for example, is already beginning to affect the way we think of hitherto motionless artefacts like buildings. If we can model motion in objects, we are likely to work hard to actually make "the real thing" move.

When we consider the public image of works of design, materialized or unexecuted projects, the technology and media used are, of course, an important component of the effectiveness of the image-making effort. The types of technology and media selected already harbour a message as to the design's general ambiance: "high-tech" buildings, for example, would look a bit ridiculous if represented publicly in a "low-tech" technique. A state-of-the-art presentation, using the latest technological means, is usually seen favourably and has commercial advantages as it projects competence and up-to-date design skills. Indeed, in today's marketplace advanced representational skills are a valuable asset that is generously rewarded by employers and clients. The public images transmitted by designs and designers are thus not only a cultural matter, an agent for education, influence, and transformation. They are also the single most important ingredient in the kind of public relations that is the threshold requirement for survival in our media-driven culture.

In Conclusion

The central role played by representation in designing is captured in the following quote from the opening statement in *Retrospecta*, an annual (student-edited) publication of the Yale School of Architecture that presents work done in the school during the previous academic year:

Projects develop through sketches in cardboard and on trace [paper]; they are pushed further through exacting CNC-milled projects and detailed renderings. But students are as likely to work through complex details by hand and to look to the computer as a means to produce quick analytical sketches.

(Paradiso et al. 2002, p. 2)

The students' statement reflects more than today's technological state of the art. It also hints at the criticalness of choosing the most appropriate representational means for every phase of any given design task. We hope to have demonstrated the pervasiveness of representation, in particular graphic representation, in designing and in a rapidly expanding design culture, in terms of cognition, history and culture, and technology and media. Representation is part and parcel of the designer's thinking and reasoning, individually and collaboratively. It encapsulates the designer's repertoire and personal preferences, despite its dependence on available technology and the ownership of skills. In the public realm, representation is used to communicate more than design facts – it conveys messages concerning a wide cultural, social, and economical context in which the design has been conceived and is to be interpreted. Taken together, these parameters describe a possible epistemological basis for the study of design representation.

Acknowledgement

The author wishes to gratefully acknowledge receipt of grant #022–740 from the Fund for the Promotion of Research at the Technion, which partially supported the writing of this chapter. A first version (here extensively revised) was published in French: "*Processus privé et image publique dans la représentation architecturale.*" In: *Les Cahiers de la Recherche Architecturale et Urbaine*, 8(13–22) 2001.

Note

1. The industrial fabrication of translucent "sketch paper" did not begin until the second or third decade of the 19th century. The first photograph was taken at about the same time.

References

Alexander, C 1964. Notes on the synthesis of form. Cambridge, MA: Harvard University Press.

Benaïssa, A and F Pousin 1999. Figuration et négociation dans le projet urbain. Les cahiers de la recherche architecturale et urbaine, no. 2–3, 119–134.

Bugaisen, N 2000. Conventions and personal preferences in graphic architectural representation: Historical evolution and transitions in the modern era. Masters thesis, Technion.

Eisenman, P 1986. Houses of cards. Oxford: Oxford University Press.

Evans, R 1995. The projective cast. Cambridge, MA: MIT Press.

Fish, J (this volume). Cognitive catalysis: Sketches for a time-lagged brain. In Design representation, edited by G Goldschmidt and WL Porter.

Fish, J and S Scrivener 1990. Amplifying the mind's eye: Sketching and visual cognition. Leonardo 23:117–126.

Goldschmidt, G 1991. The dialectics of sketching. Creativity Research Journal, 4(2):123–143.

—— 2002. Read-Write acts of drawing. In TRACEY, website dedicated to contemporary drawing issues; issue on Syntax of Mark and Gesture. http://www.lboro.ac.uk/departments/ac/tracey/somag/gabi.html Loughborough University, UK.

Goldschmidt, G and E Klevitsky (this volume). Stirling's museum projects in Germany: Graphic representation as reconstructive memory. In Design representation, edited by G Goldschmidt and WL Porter.

Grignon, M 2000. Deux brouillons: Le croquis et la maquette. In Genesis No. 14: Architecture, edited by P-M de Biasi and R Legault. Montreal: Centre Canadien d'Architecture/Jean-Michel Place, pp 153–162.

Grumbach, A 1979. Personal communication. (Antoine Grumbach is a French architect.)

Heath, T 1991. What, if anything, is an architect? Melbourne: Architecture Media Australia.

Herbert, DM 1988. Study drawings in architectural design: Their properties as a graphic medium. Journal of Architectural Education 41(2):26–38.

Klevitsky, E 1997. Three dimensional composition in architectural education in the 20's: The Bauhaus and Vkhutemas Schools. Masters thesis, Technion.

Klotz, H 1988. The history of postmodern architecture. Cambridge, MA: MIT Press.

Paradiso, A, E Baxter, and M Baumberger, eds. 2002. Retrospecta 01–02. New Haven: Yale School of Architecture.

Quantrill, M 1983. Alvar Aalto: A critical study. London: Secker & Warburg.

Quarante, D 1994. Eléments de design industriel. Paris: Polytechnica.

Schank-Smith, K 2000. L'esquisse et l'intervalle de la création. In Genesis no. 14: Architecture, edited by P-M de Biasi and R Legault. Montreal: Centre Canadien d'Architecture/Jean-Michel Place, pp 165–176.

Schön, DA 1983. The reflective practitioner. New York: Basic Books.

Schön, DA and G Wiggins 1992. Kinds of seeing and their function in designing. Design Studies 13(2):135–156.

Suwa, M, JS Gero, and T Purcell 1999. Unexpected discoveries and S-inventions of design requirements: A key to creative designs. In Computational models of creative design IV, edited by JS Gero and M-L Maher. Sydney: Key Centre of Design Computing and Cognition, University of Sidney.

Tapert, A 1990. SwidPowell: Objects by architects. New York: Rizzoli.

Trevisiol, R 1982. The round house. In La Casa Rotonda, edited by M. Botta, 81–83. Milano: L'erba Voglio.

Verstijnen, IM, JM Hennessey, C van Leeuwen, R Hamelm, and G Goldschmidt. 1998. Sketching and creative discovery. Design Studies 19(4):519–546.

Woolley, M (this volume). The thoughtful mark maker – representational design skills in the post information age. In Design representation, edited by G Goldschmidt and WL Porter.

Wrede, S 1986. Mario Botta. New York: Museum of Modern Art.

Yates, P (this volume). Distance and depth. In Design representation, edited by G Goldschmidt and WL Porter.

Index

Note: Page numbers in *italic* type refer to figures, while **bold** type indicates a major entry.

Aalto, Alvar xiv, 208–9
abstract 86–7
abstract representation 87, 90–1
accuracy 190

Ackermann, Edith 68
action
  and artefacts, observing and experiencing 145
  and pragmatics 141–2
  categories selected from videotapes *133*
  categorizing 131–2
  design representation through 127–48
  kinds of actions 67
  locating/indicating 141, 144
  model 145–6, *145–6*
  triggers 144
  value to talk 140
action-types 144
  number of events 137–8
Aesthetics and History 32
affordances 96
Alberti 211
Alberti, Leon Battista 4, 6, 16
Alexander, Christopher 208
All-Terrain Vehicle 88, *88*, 98–101, *100*
along the way of design 63–6, 78
aluminum crane 88, *88*, *91*
ambiguity *51*, *53*, 168
analysis of design process 111
analytical drawings 55
Anderson, Stanford 53
Archer, Bruce 180
architectural design xii, xvi
architectural perspectives
  designers' objects 63–79
  distance and depth in 3–35
  graphical representation 37–61
architectural representation xii, xiii, 60

Architectural Representation and the Perspective Hinge 6
architecture xii, xvi, 42–4, 46–7, 50, 53–4, 56–60, 203, 208–9, 211–12, 216
  democratizing 56
Architecture in the Twentieth Century 47
Arnheim, Rudolf 51, 57
"artefacting" 128
artefacts xiii, xiv, xvi, 127, 127–48, 142, 205, 207, 215
  and action, observing and experiencing 145
  in talk and action events 142
  model 145–6, *145–6*
  team members using 142
Attneave, F 51
augmented reality 193, 199
availability of information 105, 115–16, 122, *123*
axonometric drawings 37–9, *40–1*, 44, *44–5*, 47, 51, 55–7, 58, 213–14
axonometric views xiii, 214

Bacardi Administration Building, Mexico City *15*
backrest demonstration *136*
Baddeley, AD 158
Bailetti, AJ 146
Bamberger, J 207
Baudrillard, Jean 3
Bauhaus 211
Berenson, Bernhard 32
Berger, Ferdinand *14*
Black, Max 168
Blackmore, S 163
bodily learning 94
Botta, Mario 209, *210*
bottle of wine 69–70, *69–70*
brain 151, 153–8, 160, 163, 165, 167–9, 171–2, 181
  evolution 155

representation within 167–8
  translation mechanisms 153
brain-imaging studies 156, 168
Brereton, Margot F 85, 88, 90, 93
Brooks, LR 165
Brunelleschi 6, 211
Bruner, Jerome S 159
Bugaisen, N 209

Calvin, WH 157
candle + glass 70, *70*
Carpenter Center 71, *73*
Castelvecchio, Verona *17–18*
Chaiklin, S 83
Chambers, D 51
Chandigargh *24*
characteristics (categories) of the communication 108
Chase, WG 173
chemical catalysis 169–70, *170*
Chomsky, Noam 156
Cleveland House *31*
Clottes, J 160
Cocude, M 169
cognition xvii, 204–5, 216
cognition in design representation 204–8
cognitive analysis 179
cognitive aspects of public image 207
cognitive catalysis 151–84
cognitive science 57
cognitive translation catalyst, sketch as 174, *174*
collaboration 114, 116
collage 37, 41, 51–5, 58
Collage City 53–4
co-location 25, 27, 31
Colquhoun, Alan 51
communication xiv, xv, xvi, 67, 108–9, 111, 113–17, *114*, 120–3, *123*, 124, *124*, 204–5, 207
  in successful solution search situations 122–4, *123*

computational support tools 215
computer-aided design (CAD) 187, 190, 215
computer-aided representation in transition 197–8, *198*
computer aided design (CAD) 215
computer applications 215
computer software design applications 199
computers 185, 187, 189
conceptual art 57, 213
conceptual drawings 38
concrete 86–7
continuous data-processing 189
continuous design data-processing 200
continuous representational data flow 191
corporate identity 208
Coyne, Richard 191
craft 197
    expression through 74–5, *74*
Crafts Centre, Balerna, Switzerland *210*
Cranbrook Institute of Science, Bloomfield Hills *32*
Crawford, Alan 175
creative serendipity 188
creativity 190, 208
critical situations xv, 105, 112, *112*, 113, *113*, 115–17, 120–2, *120*–3, 124
Croset, Pierre-Alain 4
cross-discipline team 128
cultural continuum 165–6
cultural evolution 162–5
culture xvi, xvii, 204, 208, 216
customization 196

data analysis, videotapes 131
data gathering 87–8, 130
Dawkins, Richard 163
Delaney, PF 171
Delft Protocol 128
Denis, M 169, 173
Dennett, Daniel 163
depiction 159, 165, 167–8, 181
depth 3, 8–18, 20–3, 26, 28–30, 33
Descartes, René 5
description 165–6, 168, 171, 181
descriptive–depictive continuum 167
descriptive–depictive tree 172
descriptive information 167
design xi–xvii, 37–8, 41–3, *43*, *45*, 46–51, 53, 55–6, 58–9, 151, 153–5, 162, 164, 166–70, 172, 175–80, 203–5, 207–12, 214–16
design concepts xiii, 46–7, 49
design culture 216

design inquiry 64, 68
design practice 191
design practitioners 199
design problems 204
design process xii–xv, 39, 41, 43, 55, 57, 190, 206, 215
design profession 199
design proposals 90
design representation xi, xiii, xiv, xvii, xviii, 161–2, 173, 189, 203–4, 210, 214, 216
    abstract vs. concrete 85–7
    as reconstructive memory 58
    cognition in 204–8
    external 205
    futures 193–200
    internal vs. external 84–5
    post-information age 185–201
    private process 207–17
    product development 105–26
    public image 207–17
    self-generated vs. ready-made 85
    through action 127–48
    through talk 127–48
    transient vs. durable 85
    types 203–4
design skills 186
design teams xiv
design thinking xvi
design thinking research xi, xii, 57
designer 189
designers' objects 63, *63*–79, *65*, *69*, *78*
    bottle of wine 69–70, *69*–70
    candle + glass 70, *70*
    Carpenter Center 71, *73*
    discourse with objects 77–8
    examples 66
    expression through craft 74–5, *74*
    kinds of objects 68
    made during design 64–5
    metaphor of light 76, *77*
    metaphor of the book 75, *75*
    narrative objects 71
    objects of objects 69–70, *69*–70
    replication 71
    synthetic approach to reading 74–7, *74*–7
    unlikely juxtaposition 75–6, *76*
deskilling xvii, 187, 215
desktop publishing 189
Dewey, John 66
dimensions 84
distance 3, 5, 8–9, 11, 13–14, 16–18, 21, 33
distributed cognition 83, 83–103
DNA sequences 156

doing tools 193
Dourish, P 84
Downcast Eyes: The Denigration of Vision in Twentieth-Century French Thought 5
drawing as training for the mind 180
drawings xiv, xv, 37–9, 41, 43, 47, 50–1, 53–8, 60, 151, 205, 207, 209, 211, 213–14
durable representation 85

Ecole des Beaux Arts 211
education 178–9
Eisenman, Peter *212*, 214
enfilade 17–20, 25, 27–8
engineering xii, xiv–xvi
engineering concepts xv
engineering design xiv, xvi
engineering fundamentals 87
engineering perceptions, critical situations of product development 105–26
engineering perspectives
    artefacting 127–48
    distributed cognition 83–103
    prototyping 127–48
Ericsson, K Anders 171
Evans, Robin 14, 16, 56–7, 211
evolution 153–6, 162–4, 180
    by natural selection 153
evolved skills 192
exchange of information 105
Exeter Library at Phillips Academy 74
exploration 64, 66
expression through craft 74–5, *74*
external representation 83–4, 87

"fan motif" 208
feedback 145, 189
"feedthrough" 145
Feynman, RP 102
figure and ground 50–1, 53, *53*, 55
Fish, J 205
Francesca, Piero della 22, *23*
front-edge phase of designing 56
fruit press 107, *107*
fundamental concepts 87
future strategies 198–200

gaps 101
generality 168
German museum projects 37–61
German Pavilion, Barcelona 14, *15*
Gestalt psychology 207
gestures xiv

Gibson, James 179
Giedion, Siegfried 8
Goldschmidt, Gabriela 175
Gombrich, Ernest 161
Goodman, Nelson 67
Goodwin, C 83
Görner, R 115
Gössel, Peter 47
graphic representation xiv, 37–61
Grotte de Cougnac, France 161, *161*
Grotte des Trois Frères, France *162*
ground and figure *see* figure and ground
group climate 116, 121–3, 125
group organization 123
group performance 116, 125 effectiveness of 116
group prerequisites in product development 111
Gwathmey, Charles *213*, 214

habitual skills 187
Hadid 60, 214
'hands-on' skills 186
hard line drawings 205
hardware 83, 87
   as chameleon 97–8
   as medium of integration 98–101
   as starting point 93–6, *96*
   as thinking prop 95–7, *97*
   developing repertoire through integration 98–101, *100*
   negotiating between abstract and material representations 90–2, *90–2*
   roles in learning 92–3
   roles in mediating design negotiations and associated learning outcomes 94
hardware repertoires xv, 84, 90, 98
hardware starting points 93
hardware thinking props 95
Harrison, P 127
Harrison, S 83, 93, 127–8, 132, 144
Hayes, JR 173
Heath, Tom 203
high-order skills 187
high-performance studio *196*
Hitch, VJ 158
Hoesli, Bernhard 9, *10–11*, 11
Hoffman, DD 171
Holl, Steven 5, 28, 30, *31–2*
Houses of Cards *212*, 214
human factors 207
Human Problem Solving 57
Hutchins, E 83
hybrid imagery 173

ill-structures problems xv
imagery 152–3, 157, 160, 162–3, 165, 167–8, 173–4, 178–80, 205
implicit knowledge retrieval 170–2
impromptu prototyping xvi
indexical 97
industrial design xii
Infinite Corridor *72*
information representation, methods 106–8
information search 116–21
   in critical situations *120*
   types of 120
information systems, use in daily work 114, *114*
information technology xvii, 198
   training 196
information transfer 105, 112–13, 115, *115*, 116–17, *117*, 124, *124*
   categories of 117
   categorised dialogue 117–18
   in critical situations 112–21, *112–15, 120*
   role in design process 112
innovation 46, 58, 208, 214
integration, hardware as medium of 98–101
intelligent interactive pencil 199
intelligent marker 198–200
intelligent systems 189
interaction analysis 89
intermediate products 188
internal representation 84
ISN 156
isocephaly 13–14

Jameson, Fredric 3
Jay, Martin 5, 12
Jeanneret, Pierre 29
Jencks, Charles 45–6
Jerison, HJ 157
Jones, E. 43

Kahn, Louis *12–13*, 18, *19–20*, 66, 74
Karnak 66, *66*
Kepes, Gyorgy 8, 11
kinaesthetic memory triggers 93–5
kit of tools 209
kitchen scales 88, *88*, 93
Klevitsky, E. 211
Klotz, H 46–7
Koetter, Fred 53
Kosslyn, Stephen M 160, 166, 169, 173
Krier, Leon 43, 55

La Tourette 3–4, 28
language xiv, 151, 153, 156, 158, 162, 164–8, 180

language instinct 155–8
Lave, J 83
Lawson, Bryan 151
Le Calle Saint Cloud *24*
Le Corbusier 3, 6, 9, *9–11*, 16, 19–20, *21*, 22–6, *22, 24, 26*, 27–30, *29–30*, 33, *42*, 47, 49, 66, 71, *207*
League of Nations 29–30
learning by doing xiv
learning process xv
Lefaivre, L 45–6, 55
Leifer, LJ 90
Leonardo da Vinci 161, 206, *206*
Leuthäuser, Gabriele 47
Levallois technique 155
Levin, David Michael 9
Lewis-Williams, D 160
Libeskind 60, 214
life-long learning 197
Litva, PE 146
Logan, Gilbert D 128
long-term memory 155, 159, 167–8, 171–2
long-term visual memory 170, 178
Lotus International 51
Luria, Alexander 178

Macann, C 84
McCullough, M 186
Maison La Roche-Jeanneret 25–6
manipulation 198
manual dexterity 185
marker-pen systems 190
Marr, David 171
Martinelli, Antonio 27
material representation 83, 87, 90–1
meaning 63–4, 74
media xvii, 204, 211, 214–16
memes, sketches as 162–5
memory of the site 213
mental catalysis 169–70, *170*
mental curve tracing 172
mental images xvii
mental translation catalyst 174
Merleau-Ponty, Maurice 6, 8–12, 17, 25, 28–30, 33
metaphor of light 76, *77*
metaphor of the book 75, *75*
Mies van der Rohe, Ludwig xiv, *15*, 16
Mikveh Israel Congregation *13*
Miller, CM 99
mind–body dualism 5
Minneman, S 83, 93, 127–8, 132, 144
model 145, *145–6*
modernism 42, 45–7, 211, 213
Moll, LC 84

Molnar, Farkas 7
Morteo, Enrico 50
Museum of Contemporary Art, Helsinki 31

narrative objects 71
National Curriculum 180
National Gallery, London 164, 164
Necker Cube 51
negative evaluations in critical situations 120-1, 121
negotiation 85, 90
negotiation process 90-2, 90-2
Neuen Schauspielhaus 13, 14
Neuilly sur Seine 30
Newell, Allan 57, 168
Nordrheine-Westfalen Museum 37, 38, 44, 49, 59
    principal design concept 48-9
Norman, D 199

object xi, xiii, xiv, xvi, 3-6, 8-12, 14, 16, 18, 20, 25-6, 28, 30-1, 33
O'Conaill, B 143
Oeuvre Complète 25
100 Contemporary Architects: Drawings and Sketches 47
origination 198
ownership of skills 192

Palace of the League of Nations 29
palaeolithic 153-5, 157, 159-60, 162, 177
Palmer, Stephen 165
Panofsky, Erwin 6
parallax 28, 30-1
parametric systems 190
particleboard production plant 107, 107
Pelletier, Louise 6, 8
perception 151, 159-60, 179
Pérez-Gómez, Alberto 6, 8
performance 123
personal design technology audits 196
perspective 3, 6, 11-13, 16-20, 23, 25, 27, 30-1
Perspective as Symbolic Form 6
phenomenal transparency 8-9
phenomenology 84
Philadelphia Midtown Civic Center Development 12
pictorial design narrative 50
pictorial narrative xiii
Pinker, Steven 156
Plante, S 187
Plaster Cast Gallery 27, 27
Poincaré, H 168
Porter, William L 58
positive statements in critical situations 121, 122
Post-Information Age xvii, 185

post-Modernism 45-7
Postman, L 159
postmodern representation 45-6
postmodernism xiii, 58, 211, 213
pre-industrial skills 186
prescribed or regulated skills 191-2
presentation drawings 205
primary isomorphism 166
private design enterprise 215
private design search process 211
product development
    compiling design process data using direct and indirect investigation methods 109, 109
    critical situations 105-26
    external conditions and design process 108-9
    factors influencing design process 106, 106
    group prerequisites in 111
    individual prerequisites 109-10
professional development 193-7
promenade architecturale 48-9, 56
prototypes 83
prototyping 127-48
public image
    cognitive aspects of 207
    of works of design 215
purposeful skills 199
purposeless skills 199

qualitative analysis 89

Radcliffe, David F 127-8
Raphael 210
ready-made hardware 85
recognition theory 167
reconstructive memory 37, 37-61, 58
    design representation as 58
Reddy, MJ 77
reflective conversation 83, 91-2
'reflective practitioner' 190, 195
rehabilitation engineering 128
Rehabilitation Engineering Centre (REC) 128
rehabilitation engineering team 128-9, 129, 134
Reisberg, D 51
reiteration-representational tool feedback loops 189
Renaissance 43, 53, 58, 214
representation xi, xii, xiii, xv-xviii, 37-9, 41, 43-7, 50-1, 54-6, 58, 64, 66, 68, 78, 152, 159-60, 165-9, 173, 176, 179-81, 214-16

representation technologies 197
representational design skills 185
representational skills 191
representational technologies 187
representations xiv, 203-5, 207-11
rerepresentation 84
Richards, M 171
Rowe, Colin 3, 4, 8, 11, 28-30, 33, 42, 53

Salk Institute, La Jolla, California 18, 19-20
Scarpa, Carlo 17-18, 27, 27, 28
Schinkel, Karl Freidrich 13, 14, 42, 46, 48, 53, 60
Schön, Donald A 83, 91, 207
Schumacher, Thomas 22
seating clinic 128-9, 129, 132, 134-5
secondary isomorphism 166
self-directed skills 192
self-generated representations 85
self-originated skills 192
serendipity 190
shaman theory 160
Shepard, Roger N 166
short-term memory 171
shoulder strap 143
Simon, Herbeert A 57, 168, 173
sketch representation 190
sketches xiii-xv, 37, 39, 43, 47, 51-2, 55-7, 59-60, 151-3, 158, 161, 163-4, 167, 169-74, 174, 175-6, 179, 205-6, 206, 207, 209, 214, 216
    as cognitive translation catalyst 174, 174
    as memes 162-5
    functions of 151-3, 152
    use in design cognition 151-84
sketching xvi, xvii, 205-7, 209, 214-15
    as mental translation 169-72
    role of 205-7
sketching tools 189-91
skill-as-creative stimulus 194
skill-based tools 194
skills xvii, 190
    acknowledgement of 186
    control of 191-3
    development of 191-3
    division between professional (white collar) and non-professional (blue collar) 187
    evolving ownership 192

skills (*continued*)
  in overcoming
    imperfections or
    constraints 190
  keeping ahead
    professionally, socially
    and psychologically
    187–9
  ownership 215
  past, present and future
    185–201
  prescribed or regulated
    191–2
  theoretical analysis 187
  through history 185
skin rigourism 55
Slattery, P 127
Sloman, A 166
Slutzky, Robert 8, 11, 29–30
social constructivist 84
social relations xiv
software design applications
  199
solution search 116–17, *118*,
  119–23, *124*, 125
Space, Time and Architecture 8
spatial images, manipulating
  and scanning 172–5
spatial relations xiii
Staatsgalerie, Stuttgart 37, *39*,
  45, 48, 50–1, *51*, *52*, 53, 59
Standing, LG 170
Stirling and Wilford 213
Stirling, James 37, 37–61, *38–9*,
  41–2, 44–8, 50, 52–60,
  213, 214
  profile 42–4
Stirling, James and Partners 7
Stretto House, Dallas *32*
student engineering designers
  xv
Sturt, G 197
style 208
subject 4–6, 8–13, 18–19, 23,
  25–7, 30, 33
subject–object 14
supervisory central executive
  168
synergistic ecphory 172
'systematic play' 195

talk
  and action events 144
  categories that participants
    use with design 134
  categorizing 132–4, *133*
  design representation
    through 127–48
  model 145–6, *145–6*
  value of action to 140
talk-types 144
  high-specificity 141
  number of events 137
Tang, JC 127
Tartar, D 132
team xiv, xv
teamwork xv, 108, 207–8
technology xvii, 187, *188*, 204,
  214–16
  and design 185
  development 214–16
Tesla, N 102
The Architectural Review 57
The Flagellation of Christ 22,
  23
*The History of Postmodern
  Architecture* 46
The Red Cube *7*
Thomson, EP 185
three-dimensional xiii
three-dimensional abstract
  compositions 211
three-dimensional models xii
tool making 188
tool outcomes 188
tool use 188
tools 188
traditional architecture 208
transient representation 85
transient skills, transformation
  into long-term skills
  199–200, *200*
travel sketches 66, *66*
Tschumi 60, 214
Tulving, E 172
Tzonis, Alexander 45–6, 55

ubiquitous computing 193–7
Ullman, S 166
ultimate design 64–5, 78
uncertainty, degrees of 168

vagueness 168
Venturi, Robert 164, *164*
verbal communication 114–15
verbal design representations,
  initiation 115–16
Video Interaction Analysis
  (VIA) 89–90
video recordings 130
videotapes
  action categories selected
    from *133*
  categorizing action 131–4,
    133
  categorizing talk 132–4, *133*
  data analysis 131
  sample exploratory analysis
    134–42
viewing tools 193
Villa La Roche *26*
Villa Meyer *30*
Villa Savoye *21–2*
Villa Stein 9, *10–11*, 25
virtual reality 215
Visual Thinking 57
visualizing instinct 155, 158–62
vocabulary 144

Wallraf-Richartz Museum *7*,
  37, *40–1*, 59
Weiser, Mark 194
Welch, Robert 175–6, *176–7*
*What, if Anything, is an
  Architect?* 203
Whittaker, S 143
Wiesbaden *54*, 55
Wilford, Michael 56, 59–60,
  213
Williams, George 153
Wittgenstein, Ludwig 78
Woolley, Martin 215
word-by-word transcription
  108
working memory 152–3, 158,
  160–2, 165, 168–9, 171–3,
  178–9
worlds 65, 69, 75, 78

Yamazaki Serving Collection
  175–8, *176–7*
Yi-Luen Do, Ellen 194